Water Sources

PRINCIPLES AND PRACTICES
OF WATER SUPPLY OPERATIONS SERIES

Series Editor, Harry Von Huben

Water Sources, Second Edition

Water Treatment, Second Edition

Water Transmission and Distribution, Second Edition

Water Quality, Second Edition

Basic Science Concepts and Applications, Second Edition

———————————————

Water Sources

Second Edition

American Water Works Association

Water Sources, Second Edition
Principles and Practices of Water Supply Operations Series

Library of Congress Cataloging-in-Publication Data

Water sources — 2nd ed.
    xii, 202 p., 18x23 cm. — (Principles and practices of water supply operations series; [1])
    New ed. of: Introduction to water sources and transmission. 1985.
    Includes bibliographical references and index.
    ISBN 0-89867-778-5 (hard cover)
    1. Water supply.   2. Water quality.   I. American Water Works Association
II. Introduction to water sources and transmission.   III. Series.
TD390.W38   1995
628.1'62—dc20   95-22247
                                                                         CIP

Disclaimer

Many of the photographs and illustrative drawings that appear in this book have been furnished through the courtesy of various product distributors and manufacturers. Any mention of trade names, commercial products, or services does not constitute endorsement or recommendation for use by the American Water Works Association or the US Environmental Protection Agency.

ISBN 0-89867-778-5

Cover and book design by Susan DeSantis
Editor: Phillip Murray

Printed in the United States of America

American Water Works Association
6666 West Quincy Avenue
Denver, CO 80235
(303) 794-7711

CONTENTS

FOREWORD

*W**ater Sources* is part one in a five-part series titled Principles and Practices of Water Supply Operations. It contains information on water supply sources, the use and conservation of water, source water quality, and source protection.

The other books in the series are

Water Treatment

Water Transmission and Distribution

Water Quality

Basic Science Concepts and Applications (a reference handbook)

References are made to the other books in the series where appropriate in the text.

A student workbook is available for each of the five books. These may be used by classroom students for completing assignments and reviewing questions, and as a convenient method of keeping organized notes from class lectures. The workbooks, which provide a review of important points covered in each chapter, are also useful for self-study.

An instructor guide is also available for each of the books to assist teachers of classes in water supply operations. The guides provide the instructor with additional sources of information for the subject matter of each chapter, as well as outlines of the chapters, suggested visual aids, class demonstrations where applicable, and the answers to review questions provided in the student workbooks.

The reference handbook is a companion to all four books. It contains basic reviews of mathematics, hydraulics, chemistry, and electricity needed for the problems and computations required in water supply operations. The handbook also uses examples to explain and demonstrate many specific problems.

ACKNOWLEDGMENTS

This second edition of *Water Sources* has been revised to include new technology and current water supply regulations. The material has also been reorganized for better coordination with the other books in the series. The authors of the revision are R. Patrick Grady, Assistant Editor, and Peter C. Karalekas Jr., Editor, with the *Journal of the New England Water Works Association*.

Special thanks are extended to the following individuals who provided technical review of all or portions of the second-edition outlines and manuscript:

> Tom Feeley, Red Rocks Community College, Lakewood, Colo.
> Eugene B. Golub, New England Institute of Technology, Newark, N.J.
> William T. Harding, Ingersoll-Rand Company, Scranton, Pa.
> Stephen E. Jones, Iowa State University, Ames, Iowa
> Kenneth D. Kerri, California State University, Sacramento, Calif.
> Gary B. Logsdon, Black & Veatch, Inc., Cincinnati, Ohio
> Garret P. Westerhoff, Malcolm Pirnie, Inc., White Plains, N.Y.

American Water Works Association staff who reviewed the manuscript included Robert Lamson, Ed Baruth, Bruce Elms, and Steve Posavec.

Publication of the first edition was made possible through a grant from the US Environmental Protection Agency, Office of Drinking Water, under Grant No. T900632-01. The principal authors were Joanne Kirkpatrick and Benton C. Price, under contract with VTN Colorado, Inc.

The following individuals are credited with participating in the review of the first edition: E. Elwin Arasmith, Edward W. Bailey, James O. Bryant Jr., Earle Eagle, Clifford H. Fore, James T. Harvey, William R. Hill, Jack W. Hoffbuhr, Kenneth D. Kerri, Jack E. Layne, Ralph W. Leidholdt, L.H. Lockhart, Andrew J. Piatek Jr., Fred H. Soland, Robert K. Weir, Glenn A. Wilson, Leonard E. Wrigley, and Robert L. Wubbena.

INTRODUCTION

Supplying water to the public is one of the most important services offered to society by water supply professionals. It has grown in importance and complexity throughout the late nineteenth and twentieth centuries, especially in the last two decades. With the advent of the Safe Drinking Water Act, signed by President Ford in 1974, a new era of responsibility dawned for the water utility operator.

Prior to that time, the operator's job was in large part mechanical. He or she kept the pumps running, the valves functioning, and the simple disinfection process from shutting down. The operator changed charts, washed filters, kept the buildings and grounds tidy, and handled a few basic chemicals.

Now the operator does all of those things, but with greater care and greater precision. He must still be a good mechanic, but now, if the job is to be done well, he must also have some basic knowledge of microbiology, chemistry, laboratory procedures, and public relations. The operator must be ready to explain to the community what his job consists of, why water rates seem to rise continually, and that an adequate supply of safe, palatable water is still a great bargain. The operator must reach out to the schools in the service area and teach students of all ages, both in the classroom and at the treatment plant, about the water treatment process.

Most of all the operator must constantly be aware of public health issues. Any action she takes must be taken only if the public health has been considered. The operator has the potential to create serious health hazards by negligence or to make dangerous situations harmless by using her knowledge and skill.

As is true of any quality product, the raw material is very important. The water utility operator's raw material is the source water. Every source is different, and the operator must become the most knowledgeable, best-informed person about the source or sources he or she is working with. This is true whether the source is a single well, a number of well fields, a spring, a river, or a lake.

This book will introduce the operator to the basic information related to water sources. The operator's goal should always be to take the available source water and perform the functions necessary to supply the customers with a safe, high-quality product.

CHAPTER 1

Water Supply Hydrology

Hydrology is the science dealing with the properties, distribution, and circulation of water in the atmosphere, on the earth's surface, and below the earth's surface. Of particular interest in studying water supply operations are the following:

- how water is constantly replenished by the hydrologic cycle
- how groundwater supplies are recharged and are available for public water supply use
- how surface water provides sources of supply for many public water systems
- the terms commonly used in the water supply industry to describe the volume and flow of water

The Hydrologic Cycle

The constant movement of water from the surface of the earth to the atmosphere and back again is called the water cycle or, more commonly, the hydrologic cycle. The hydrologic cycle is illustrated in Figure 1-1.

The steps involved in the hydrologic cycle are as follows:

- Water moves from earth to sky (evaporation and transpiration).
- Water vapor forms tiny droplets (condensation).
- Water falls back to earth (precipitation).
- Water on the earth penetrates the ground or runs off the surface (infiltration, percolation, surface runoff).

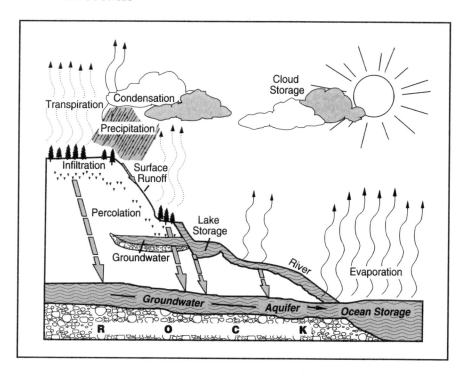

FIGURE 1-1 The hydrologic cycle

Evaporation and Transpiration

Water moves from the earth to the atmosphere in two ways. One is evaporation. Evaporation occurs when liquid water changes to an invisible gas that circulates into the air. For example, your hands become dry after washing even if you do not use a towel, and a wet driveway dries by evaporation after a rain shower. Heat from the sun speeds the evaporation process. Enormous amounts of water are evaporated from oceans, lakes, and streams, as well as from the land surface. The gas, which is called water vapor, moves through the air above the earth, and when conditions are right, the vapor condenses and clouds are formed.

Another way in which water becomes water vapor and is transferred to the air is transpiration. In this process, water from the earth is absorbed by the roots of plants such as grasses, bushes, and trees. The water moves through the plant, and the blades of grass or the leaves on the trees transfer the water to the air.

Condensation

On a hot summer day, a glass of cold liquid is soon covered with very tiny drops of water that eventually combine and run off the glass. When conditions are right, the same thing happens in the atmosphere. Water vapor condenses and forms clouds, which are large masses of very tiny drops of water or ice crystals. This change from water vapor to water drops is called condensation. Clouds can appear dazzling white or very dark and can have an endless variety of shapes, but they are always made up of water drops or ice crystals.

Precipitation

If enough droplets combine, they become too heavy to stay airborne and fall to the earth as drizzle or rain. If enough ice crystals form to become very heavy, they fall to earth as hail, sleet, or snow. This process is called precipitation. The world's freshwater supply is constantly being redistributed by precipitation.

Infiltration and Percolation

As rain or snow falls on the ground, some of it soaks into the soil. The movement of water through the soil is called infiltration. The water that soaks into the soil in the root zone of plants can be taken up by the plants to become part of the plant or again be sent into the air by transpiration.

Another portion of the infiltrated water goes back to the surface because of the capillary action of the soil. Capillary action is the force that causes water to rise above a water surface through porous soil. The capillary rise in coarse sand is only about 5 in. (125 mm), but it may be as much as 40 in. (1,000 mm) in some silty soil.

The remaining infiltrated water first replaces deficient soil moisture, and the rest continues to move downward below the root zone to a water-saturated area. This downward movement of water is called percolation.

Surface Runoff

When the surface soil can hold no more water, it is said to be saturated. Any excess precipitation then begins to flow downhill over the surface. This process is called overland runoff or surface runoff.

Most of the surface runoff then flows to a water body such as a creek, river, or lake, and much of it eventually reaches the oceans. The hydrologic cycle thus continues.

Groundwater

Groundwater is the result of the infiltration and percolation of water down to the water table through the void spaces in the soil and the cracks in consolidated rocks. (See chapter 2 for more information on groundwater sources.) An easy way to visualize groundwater is to fill a glass bowl halfway with sand, as shown in Figure 1-2. If water is poured onto the sand, it infiltrates into the sand and percolates down through the voids until it reaches the watertight (impermeable) bottom of the bowl. As more water is added, the sand becomes saturated and the water surface in the sand rises. The water in this saturated sand is equivalent to a groundwater aquifer. An aquifer is defined as any porous, water-bearing geologic formation.

When the sand in the bowl is partially saturated with water, the level of the water surface can be found by poking a hole or cutting a stream channel in the sand. As illustrated in Figure 1-3, the level of water in the channel is the same as the level of the water surface throughout the sand. This level is called the water table. The water table may also be called the top of the groundwater, the top of the zone of saturation, or the top of the aquifer.

Aquifers and Confining Beds

Groundwater occurs in two different kinds of aquifers: unconfined aquifers and confined aquifers.

Unconfined Aquifers

When the upper surface of the saturated zone is free to rise and decline, the aquifer is referred to as an unconfined aquifer, or water-table aquifer (Figure 1-4). Wells that are constructed to reach only an unconfined aquifer are usually referred to as water-table wells. The water level in these wells

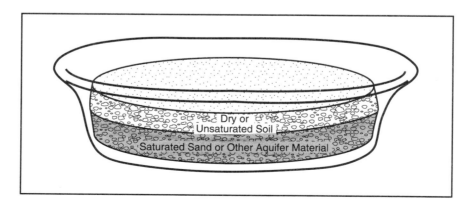

FIGURE 1-2
Simplified
example of
groundwater

Dry or Unsaturated Soil
Saturated Sand or Other Aquifer Material

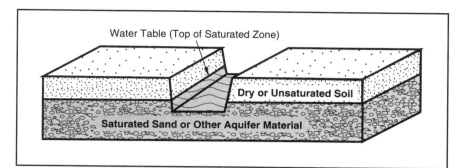

FIGURE 1-3
Illustration of a
water table

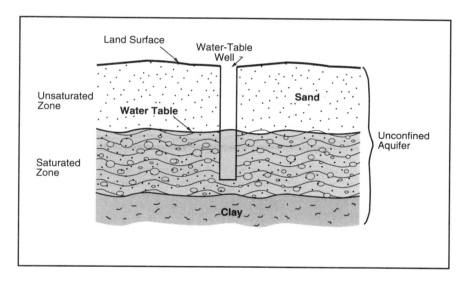

FIGURE 1-4
Cross section of an
unconfined
aquifer

indicates the level of the water table of the surrounding aquifer. The amount of water that the water-table well can produce may vary widely as the water table rises and falls in relation to the amount of rainfall.

Confined Aquifers

A confined aquifer, or artesian aquifer, is a permeable layer that is confined by an upper layer and a lower layer that have low permeability (Figure 1-5). As illustrated in the figure, the water recharge area (the location where water enters the aquifer) is at a higher elevation than the main portion of the artesian aquifer. As a result, the water is usually under pressure. Therefore, when a well is drilled through the upper confining layer, the

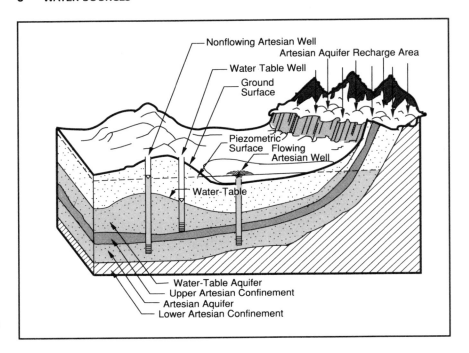

Nonflowing Artesian Well
Artesian Aquifer Recharge Area
Water Table Well
Ground Surface
Piezometric Surface
Flowing Artesian Well
Water-Table
Water-Table Aquifer
Upper Artesian Confinement
Artesian Aquifer
Lower Artesian Confinement

FIGURE 1-5
Cross section of an artesian aquifer

pressure in the aquifer forces water to rise up above the confining layer. In some cases, the water will rise above the ground surface, producing a flowing artesian well. In most cases, it will not rise as high as the ground surface and will produce a nonflowing artesian well.

A confining bed is a layer of material, such as consolidated rock or clay, that has very low permeability and restricts the movement of groundwater either into or out of adjacent aquifers.

The height to which water will rise in wells located in an artesian aquifer is called the piezometric surface. This surface may be higher than the ground surface at some locations and lower than the ground surface at other points of the same aquifer. Further discussion of the piezometric surface is provided in the hydraulics section of *Basic Science Concepts and Applications*, part of this series.

There can be more than one aquifer beneath a particular surface area. For instance, there may be a layer of gravel, a layer of clay, another layer of sand or gravel, and another layer with low permeability. Some wells are constructed to tap several different aquifers. In locations where the water quality varies in each aquifer, a well may be constructed to draw water only from the aquifer with the best quality.

Aquifer Materials

Aquifers may be made up of a variety of materials. One characteristic that is important from a water supply standpoint is the porosity, which represents the volume of space within a given amount of the material. The porosity determines how much water the material can hold. The other important characteristic is how easily water will flow through the material, known as its permeability, or hydraulic conductivity. Both of these factors determine the amount of water an aquifer will yield. Figure 1-6 illustrates the common aquifer materials.

Aquifers composed of material with individual grains such as sand or gravel are called unconsolidated formations. The range of sizes and arrangement of the grains in the material are important. For example, two aquifers — one composed of fine sand and one of coarse sand — may contain the same amount of water, but the water will flow faster through the coarse sand. A well in the coarse sand can therefore be pumped at a higher rate, and the pumping costs will be less.

Limestone and fractured rock formations are consolidated formations. These formations produce water from channels, fractures, and cavities in the rock. Some fractured rock formations can produce very large quantities of water.

Groundwater Movement

Water naturally moves downhill (downgradient) toward the lowest point. In the examples shown in Figures 1-2 and 1-3, the water will not move

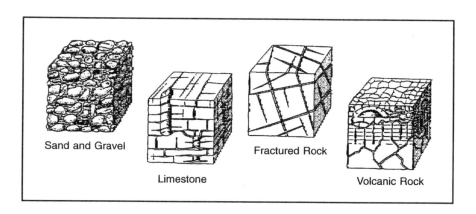

Sand and Gravel

Limestone

Fractured Rock

Volcanic Rock

FIGURE 1-6
Common aquifer materials

in any direction because the water table is flat. However, water tables are not usually flat.

Figure 1-7 shows how a water table might actually occur in nature. When rain falls on the watershed, some of the rain percolates downward to the water table. A mound of water within the aquifer is then built up above the level of the rest of the water table. The water within this mound slowly flows downgradient, increasing the level of the water table slightly and, if the water table is high enough, draining into the stream channel. In most areas, more rain occurs before the mound completely drains off, so the water table never becomes level.

The movement of groundwater is illustrated further in Figure 1-8, in which a cutaway view shows a section of the ground surface. The water table is continuous and sloped, and the groundwater moves downgradient toward the lowest point — the stream channel. In this example, the water table is above the level of the stream channel. Consequently, water will flow into the

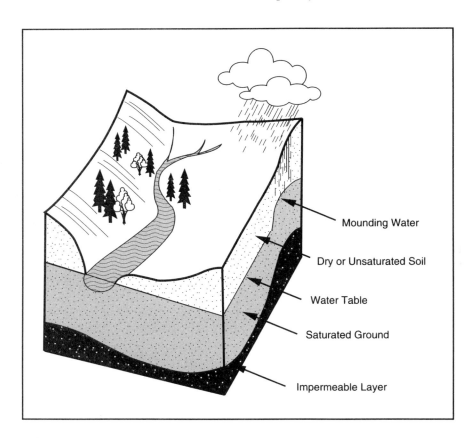

**FIGURE 1-7
Formation of
groundwater in
nature**

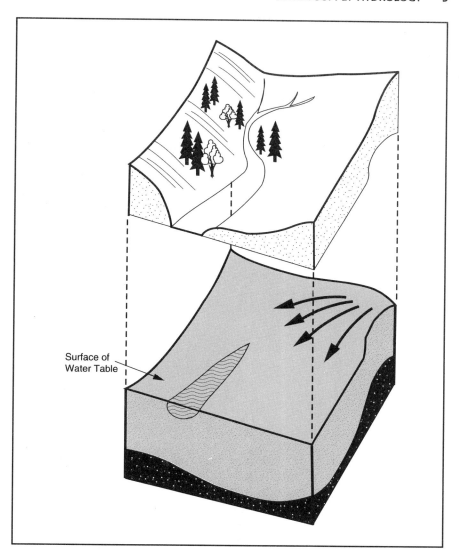

FIGURE 1-8
Groundwater
movement

Surface of
Water Table

stream. But, as explained later, if the water table is below the stream bed, water will infiltrate from the stream to the aquifer.

Springs

Springs occur if the water table intersects the surface of the ground, or if the water-bearing crevices of fractured rock come to the surface (Figure 1-9). Springs flowing from sand or soil often come from a source that is relatively

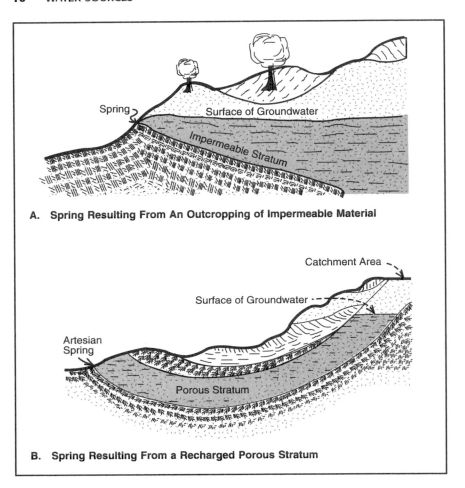

A. **Spring Resulting From An Outcropping of Impermeable Material**

B. **Spring Resulting From a Recharged Porous Stratum**

**FIGURE 1-9
Common types of
springs**

close. Springs flowing from fractured rock can come from more distant locations, and their origin is usually difficult to determine.

Because of the difficulty in determining where spring water is coming from, springs should generally be considered contaminated until proven otherwise. Many supposed "springs" have stopped flowing when a broken sanitary sewer or septic tank was repaired.

Surface Water

Surface water in lakes and streams provides a source of water for many public water supply systems. (See chapter 3 for more information on surface water sources.) The amount of surface water available varies widely by region of the country and also by season of the year.

The quality of the surface water available for water system use also varies. Some locations are fortunate to have very clean water available, whereas the only available sources in other areas require extensive treatment to make the water safe and palatable for human consumption.

Impurities in Rain and Snow

Precipitation in the form of rain or snow is the source of water for most surface water supplies. The amount of foreign material in rain and snow is minimal in comparison with the amount that the water will later have after running over or through the ground, but the water is far from chemically pure. Precipitation dissolves the gases in the atmosphere as the water falls to the ground, and it also collects dust and other solid materials suspended in the air.

The solids in the atmosphere are caused by wind-blown soil and by materials released into the air from combustion, industrial processes, and other sources, so there is considerable variation in the impurities in rain and snow. But in general, precipitation is very soft, is low in total solids and alkalinity, has a pH slightly below neutral (pH 7), and is quite corrosive to most metals. There is considerable variability depending on local conditions, but a typical analysis would yield the following attributes:

Hardness	19 mg/L as calcium carbonate ($CaCO_3$)
Calcium	16 mg/L as $CaCO_3$
Magnesium	3 mg/L as $CaCO_3$
Sodium	6 mg/L as sodium (Na)
Ammonium	0.8 mg/L as nitrogen (N)
Bicarbonate	12 mg/L as $CaCO_3$
Acidity	4 mg/L as $CaCO_3$
Chloride	9 mg/L as chlorine (Cl)
Sulfate	10 mg/L as sulfate (SO_4)
Nitrate	0.1 mg/L as N
pH	6.8

Impurities Added During Runoff

Precipitation that falls on the ground surface and does not infiltrate into the soil or evaporate into the air will travel over the land surface to a surface water body. In the process, a great variety of materials may be dissolved or taken into suspension. Surface water therefore contains impurities that reflect both the surface characteristics and the geology of the area.

The presence of soluble formations near the surface (such as gypsum, rock salt, and limestone) will have a marked effect on the chemical characteristics of surface water. On the other hand, if the geological formations are less soluble, such as sandstone or granite, the composition of the surface water in the area will tend to be more like that of rain.

If water flowing over the ground surface is exposed to decomposing organic matter in the soil, the carbon dioxide level of the water will be increased. This, in turn, will increase the amount of mineral matter dissolved by the water.

Snow as Water Storage

Snow is a very important form of water storage. Snowmelt greatly prolongs the flow in many streams. If rainfall were the only source of water for surface water bodies, many streams would have very low flows, or in some cases, no flow at all during dry periods.

Groundwater Augmentation of Surface Water

Figure 1-10 illustrates how stream flow is augmented by groundwater. If the water table adjacent to the stream is at a higher level than the water surface of the stream, water from the water table will flow to the stream. Conversely, when the water table is below the stream surface, water from the stream percolates into the ground. The recharge of groundwater from a stream is illustrated in Figure 1-11.

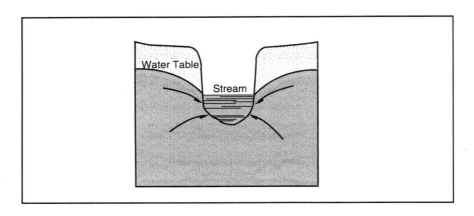

FIGURE 1-10
Surface flow from
groundwater
seepage

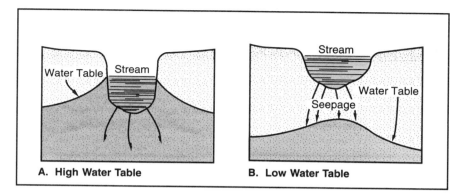

FIGURE 1-11
Stream flow
recharging
groundwater

Surface Runoff

The amount and flow rate of surface runoff are highly variable, depending on both natural conditions and human influences. In some cases, water is held on the surface for a relatively long time. This is generally good from a water resources standpoint because it allows more water to infiltrate into the ground and recharge groundwater aquifers.

If water flows off slowly, it also generally causes less erosion and creates less flooding. At the same time, the longer contact time with the soil will generally increase the mineral content of the water. Surface water running quickly off land may be expected to have all of the opposite effects.

Surface Water Collection

Surface water runoff flows along the path of least resistance. It begins to form rivulets, which then flow into brooks, creeks, rivers, lakes, and in some cases, eventually to the ocean. All water that runs off the land flows toward a primary watercourse. The land area that is sloped toward a watercourse is known as a watershed, or drainage basin. As shown in Figure 1-12, a watershed is a basin surrounded by high ground, or a divide, that separates one watershed from another.

Watercourses

Typical natural watercourses include brooks, creeks, streams, and rivers. There are also many facilities that are constructed to hasten the flow of surface water or to divert it in a direction different than it would flow under natural conditions. Ditches, channels, canals, aqueducts, conduits, tunnels, and storm sewers are some of these structures.

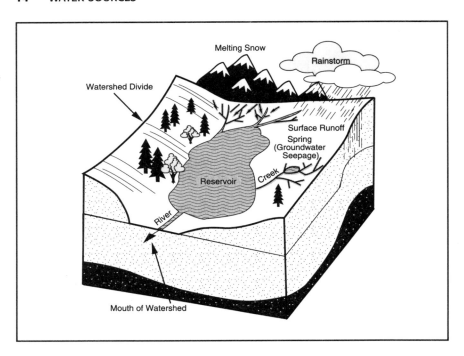

FIGURE 1-12
Schematic of a
typical watershed

Water Bodies

A water body is a water storage basin. It can be a natural basin such as a pond or lake, or a reservoir constructed to serve some specific water need. Reservoirs range from small basins constructed by farmers for watering livestock to massive reservoirs that store water for public water system and recreational use.

Volume and Flow

The branch of science that deals with fluids at rest and in motion is known as hydraulics.

Every water system has a number of measurements that must be made, recorded, and filed relating to the rate and amount of water passing through the system. Many of the instruments a water system operator uses are designed so that hydraulic calculations are done by the instrument itself and the operator simply reads the instrument.

An example of this is the water meter. Both the small residential type and the type used in a pump station or a treatment plant are designed so that the volume of water is measured continuously and the total amount of water passing through the meter is indicated in gallons (liters) or cubic feet (cubic meters). Another example is a flowmeter that converts the passage of water through it to a rate of flow. These two measurements are frequently made by the same instrument.

There are a number of terms that relate to measuring water. The following paragraphs discuss these terms.

Volume

Volume is the measurement of a quantity of water. In English units, it is most often expressed in gallons, millions of gallons, or cubic feet. In metric units, it is measured in liters, millions of liters, or cubic meters. A water volume measurement may describe such things as the amount of water contained in a reservoir or the quantity of water that has passed through a treatment plant.

Flow Rate

The measurement of the volume of water passing by a point over a period of time is called the flow rate. Following are some of the units most frequently used for expressing flow rate:

- gallons per minute (gpm) or liters per minute (L/min)
- gallons per hour (gph) or liters per hour (L/h)
- gallons per day (gpd) or liters per day (L/d)
- million gallons per day (mgd) or megaliters per day (ML/d)
- cubic feet per second (ft^3/s) or cubic meters per second (m^3/s)

Instantaneous Flow Rate

The rate at which water is passing by a point at any instant is called the instantaneous flow rate. The instantaneous flow rate through a water plant, for instance, varies constantly throughout the day.

Average Flow Rate

Dividing the total volume of water passing a point by a length of time provides an average flow rate. For instance, most water system reports compute the average use for each day in gallons per minute (liters per minute). This is obtained by dividing the total volume of water pumped for the day by 1,440 minutes in a day.

TABLE 1-1 Common measurements of water use — US customary units

Term	Definition	How to Calculate	Units of Measure	Comments on Use
Volume	Amount of water that flows through a treatment plant	Read totalizer at beginning and end of period for which volume is to be determined, and subtract first reading from second	Gallons, millions of gallons, cubic feet, acre feet	One of the basic measurements of water use
Instantaneous flow rate	The flow rate at any instant	Read directly from circular flow chart or from indicator	Gallons per minute (gpm), gallons per hour (gph), gallons per day (gpd), million gallons per day (mgd), cubic feet per second (ft^3/s)	One of the basic measurements of water use
Daily flow	Amount of water passing through plant during a single day	Measure flow volume (gallons) for a single day	gpd, mgd	Basis for calculating average daily flow and chemical designs
Average daily flow	Average of daily flows for a specified time period	Find sum of all daily flows for the period and divide by number of daily flows used	gpm, gph, gpd, mgd	Records of average daily flow rate are the basis for calculating most other measurements

Table continued next page

TABLE 1-1 Common measurements of water use — US customary units (continued)

Term	Definition	How to Calculate	Units of Measure	Comments on Use
Annual average daily flow	Average of average daily flows for a 12-month period	Average the average daily flows for all the days of the year, or divide total flow volume for the year by 365	gpm, gph, gpd, mgd	Used to predict need for system expansion
Peak-day demand	Greatest volume per day flowing through the plant for any day of the year	Look at records of daily flow rates for the year to find the peak day	gpm, gph, gpd, mgd	Determines system operation (plant output, storage) during heaviest load period — ranges from 1.5 to 3.5 times the average daily flow rate
Peak-hour demand	Greatest volume per hour flowing through the plant for any hour in the year	Determined from the chart recordings showing the continuous changes in flow rate	gpm, gph, gpd, mgd	Determines required capacity of distribution system piping — ranges from 2.0 to 7.0 times the average hourly demand for the year
Minimum-day demand	Least volume per day flowing through the plant for any day of the year	Look at records of daily flow rates for the year to find the minimum day	gpm, gph, gpd, mgd	Determines possible plant shutdown periods — ranges from 0.5 to 0.8 times the average daily flow rate

Table continued next page

TABLE 1-1 Common measurements of water use — US customary units (continued)

Term	Definition	How to Calculate	Units of Measure	Comments on Use
Filtration rate	Rate at which water is flowing through a filter	Divide the flow rate per minute flowing through the filter by the surface area of the filter	gpm/ft^2	Indicates the rate of flow through a filter, particularly to check that the design rate is not exceeded
Peak-month demand	Greatest volume passing through the plant during a calendar month	Look at records of daily flow volumes (determined from totalizer readings at the beginning and end of each month or from the sums of the daily flows for each month) to find peak month for the year	Gallons, millions of gallons, billions of gallons	Determines possible plant storage needs — ranges from 1.1 to 1.5 times the average monthly flow volume for the year
Minimum-month demand	Least volume passing through the plant during a calendar month	Look at records of monthly flow volumes (determined as for peak-month demand) to find minimum month for the year	Gallons, millions of gallons, billions of gallons	Best time of year for checking and repairing equipment — ranges from 0.75 to 0.90 times the average monthly flow volumes for a year
Minimum-hour demand	Least volume per hour flowing through the plant for any hour in the year	Determined from the chart recordings showing the continuous changes in flow rate	gpm, gph, gpd, mgd	Ranges from 0.20 to 0.75 times the average hourly demand for a year

Some other average flow rates that are frequently computed for water supply system operations are

- annual average daily flow
- peak-hour demand
- peak-day demand
- minimum-day demand
- peak-month demand
- minimum-month demand

Table 1-1 summarizes measurements of water use that are commonly used in the public water supply industry.

Selected Supplementary Readings

Basic Ground-water Hydrology. Pub. #WS 2220. 1983. US Geological Survey.

Driscoll, F.G. 1986. *Groundwater and Wells.* St. Paul, Minn.: Johnson Filtration Systems Inc.

Manual of Instruction for Water Treatment Plant Operators. 1975. Albany, N.Y.: New York State Department of Health.

Manual of Water Utility Operations. 8th ed. 1988. Austin, Texas: Texas Water Utilities Association.

van der Leeden, F., F.L. Troise, and D.K. Todd. 2nd ed. 1990. *The Water Encyclopedia.* Chelsea, Mich.: Lewis Publishers.

CHAPTER 2

Groundwater Sources

G roundwater is an extremely important resource in North America. In the United States, it represents the principal source of water for about 48 percent of the country's population, with about 95 percent of the rural population being served by wells.

Groundwater sources can be relatively simple to develop, and they often require little or no treatment before use. Small public water systems usually find groundwater the most economical source if suitable aquifers are available.

Springs and Infiltration Galleries

Springs are very rarely used as the water source for a public water supply. Ordinarily, they are not reliable sources and are difficult to protect from contamination. If a spring is to be used, its flow should be known to be reliable through the four seasons of a year and during drought years. For sanitary reasons, the collection point should be enclosed in a tamper-proof, vermin-proof concrete box, and the site should have diversion channels that route surface drainage around the collection site. Drains should be installed in the box to allow for periodic inspection and cleanout.

The purpose of an infiltration gallery is usually to collect water from a surface water source, but at a point where the flow has infiltrated through several feet of sand or gravel. In this way, most of the particulate matter is removed. As illustrated in Figure 2-1, a typical installation involves the construction of a trench that is parallel to a stream bed and is about 10 ft (3 m) beyond the high-water mark. Perforated or open joint pipe is laid in the trench in a bed of gravel and then covered with a layer of coarse sand. The upper part of the trench is filled with fairly impermeable material to reduce

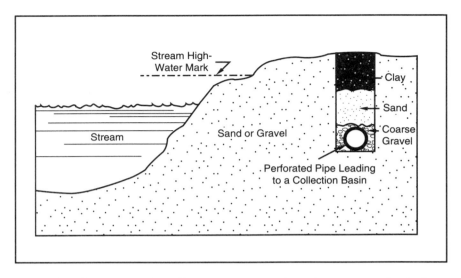

Stream High-
Water Mark

Clay

Sand

Stream

Sand or Gravel

Coarse
Gravel

Perforated Pipe Leading
to a Collection Basin

FIGURE 2-1
Example of an
infiltration gallery

entry of surface water. The collection pipes terminate in a concrete basin from which water is pumped for use.

Although the water collected in an infiltration gallery may be of much better quality than the surface source, it will not necessarily be free of all turbidity or pathogenic organisms. The water must therefore be treated in compliance with state and federal requirements for surface water.

Water Well Terminology

This section describes the parts of a well and defines common terms associated with wells.

Parts of a Well

The principal parts of a well are labeled in Figure 2-2. At the surface, a sanitary seal prevents contamination from entering the well casing. The seal has openings into the well for the discharge pipe, the pump controls, and an air vent to let air into the casing as the water level drops. A sanitary seal for a small well having a submersible pump is illustrated in Figure 2-3.

The well casing is simply a liner placed in the bore hole to prevent the walls from caving in. It is generally made of steel or plastic pipe. The well slab is a concrete area placed around the casing of some wells to support pumping equipment and to help prevent surface water from contaminating the well water.

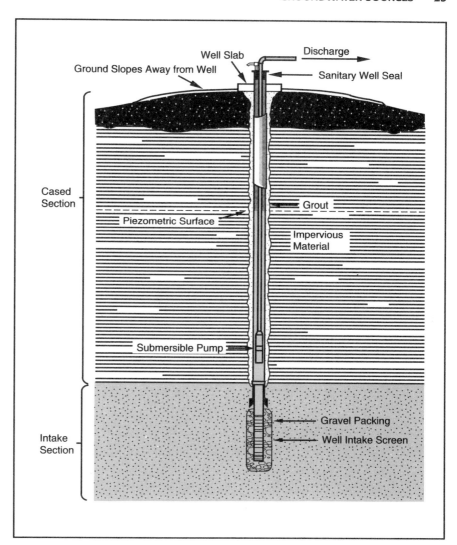

FIGURE 2-2 Parts of a typical well

Grout is cement or other material that is pumped into the space between the drilled hole and the casing to protect the casing and to prevent water from traveling along the casing from the surface or between aquifers. If the well is to draw water from sand or gravel, a well screen (intake screen) is placed in the aquifer to prevent sand from entering the well.

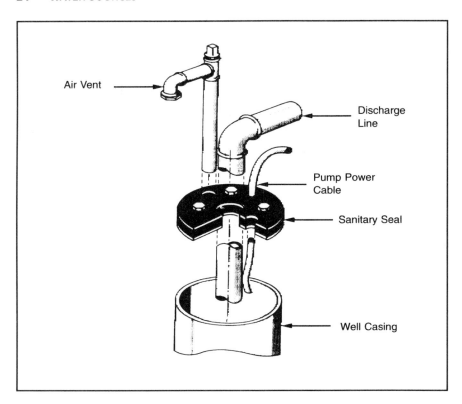

FIGURE 2-3
Components of a
sanitary seal

Well Terms

The hydraulic characteristics of a well that are important during well operation are

- static water level
- pumping water level
- drawdown
- cone of depression
- zone of influence
- residual drawdown
- well yield
- specific capacity

Many of these characteristics are illustrated in Figure 2-4.

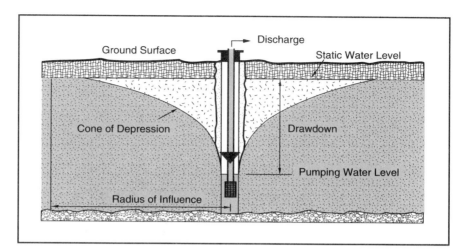

Discharge

Ground Surface

Static Water Level

Cone of Depression

Drawdown

Pumping Water Level

Radius of Influence

**FIGURE 2-4
Hydraulic
characteristics
of a well**

Static Water Level

The static water level in a well is the water surface in the well when no water is being taken from the aquifer. It is normally measured as the distance from the ground surface to the water surface. This is an important measurement because it is the basis for monitoring changes in the water table.

In some areas, the water table is just below the ground surface, but more often it is several feet below the surface. In other areas, there is no water near the surface, and a well may not pass through any water until an aquifer is intercepted at hundreds or even thousands of feet below the surface.

Pumping Water Level

When water is pumped out of a well, the water level usually drops below the level in the surrounding aquifer and eventually stabilizes at a lower level called the pumping level. The water intake or submerged pump must be located below this level.

Drawdown

The drop in water level between the static water level and the pumping water level is called the drawdown of the well.

Cone of Depression

In unconfined aquifers, there is a flow of water in the aquifer from all directions toward the well during pumping. The free water surface in the aquifer then takes the shape of an inverted cone or curved funnel called the cone of depression.

Zone of Influence

The zone affected by the drawdown extends out from the well a distance that depends on the porosity of the soil and other factors. The relative steepness of the cone of depression corresponds to a radius of influence that extends horizontally outward from the well, as shown in Figure 2-4. The length and depth of the radius define the zone of influence. If the aquifer is composed of material that transmits water easily, such as coarse sand or gravel, the cone may be almost flat and the zone of influence relatively small. If the material in the aquifer transmits water slowly, the cone will usually be quite steep and the zone of influence large.

If possible, wells should be situated far enough apart that their zones of influence do not overlap. The effect on the cones of depression of two wells located close together is illustrated in Figure 2-5. Pumping a single well reduces the water level in the other well, and pumping both simultaneously creates an undesirable interference in the drawdown profiles.

Residual Drawdown

After pumping is stopped, the water level in a well will rise toward the static water level. If the water level does not quite reach the original level, the distance it falls short is called the residual drawdown.

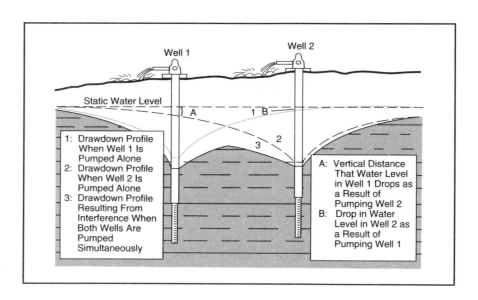

FIGURE 2-5 Overlapping cones of depression of two wells

Well Yield

Well yield is the rate of water withdrawal that a well can supply over a long period of time. The yield of small wells is usually measured in gallons per minute (liters per minute) or gallons per hour (liters per hour). For large wells, it may be measured in cubic feet per second (cubic meters per second).

When the pumpage from an aquifer continually exceeds the water recharge to the aquifer, the drawdown will gradually extend to greater depths, and the safe well yield will be reduced. If, under prolonged pumping, the drawdown extends to below the pump suction, the pump will begin to suck air. This can damage the pump, so it must be avoided.

When maximum drawdown is almost below the pump suction, pump damage can be avoided by lowering the pump setting, if this is possible. The other choice is to operate the well for only short periods of time to limit the drawdown and to allow the well to "rest" while the water level recovers before the pump is operated again.

Specific Capacity

One of the most important concepts in well operation and testing is specific capacity. It is a measure of well yield per unit of drawdown, or

$$specific \ capacity = well \ yield \div drawdown$$

For example, if the well yield is 200 gpm and the drawdown is measured to be 25 ft, the specific capacity is

$$200 \div 25 = specific \ capacity$$

$$specific \ capacity = 8 \ gpm \ per \ ft \ of \ drawdown$$

The calculation is simple and should be made frequently in the monitoring of well operation. A sudden drop in specific capacity indicates trouble such as pump wear, screen plugging, or other problems that can be serious. These problems should be identified and corrected as soon as possible.

Well Location

A well should be located to produce the maximum yield possible while still being protected as much as possible from contamination. A number of methods can be used for determining likely locations for installing a new well.

Existing Data

When a new well is to be constructed, the first places to look for information are the state and federal geological or water resources agencies. In most states, data from previous well logs have been used to draw maps of the geology of various regions and to define water-bearing layers. The agencies should also be of assistance in identifying the productivity of aquifers in the region and determining which of them is capable of providing flow to additional wells.

They will probably also have information on water quality, including data on hardness, iron and manganese, sulfur, nitrate, radionuclides, and other water characteristics that might be of concern. The state agency may also be able to provide some information on the possibility of contamination of the aquifer from organic chemicals.

The owners of surrounding public and private wells are another source of information. If there is more than one aquifer in the area, it is important that the information obtained applies only to wells in the aquifer being considered for the new well. Although there are often some differences in water quality within an aquifer, the information from existing wells gives a fair idea of what can be expected in a new well.

Local well-drilling contractors are the third source of information that should be used. They usually have extensive practical information on where wells can be developed and on the quantity and quality of water that will be obtained.

Likely Locations

As one might expect, groundwater is likely to be present in larger quantities under valleys than under hills. Valley soils containing permeable material washed down from mountains are usually productive aquifers. In some areas of the country, the only groundwater of usable quality is found in river valleys. Coastal terraces as well as coastal and river plains may also have good aquifers.

Any evidence of surface water, such as streams, springs, seeps, swamps, or lakes, is a good indicator that there is groundwater present, though not necessarily in usable quantities. Sometimes vegetation will show the location of groundwater; an unusually thick overgrowth may indicate shallow groundwater.

Exploration

After all existing data and likely locations have been examined, there still may not be enough information to risk spending a large amount of money on a full-size well. It is then advisable to conduct an underground exploration to provide additional information on geologic formations and aquifers. Some of the more common methods include using seismic and resistivity tests, drilling test wells, and using computer modeling.

Seismic and Resistivity Tests

If there is inadequate information on groundwater in an area, the next step in exploration is to employ a professional firm to run seismic and resistivity tests. The seismic test measures the speed at which a shock wave travels through the earth. It is used to identify geologic formations that may contain groundwater. A small charge of dynamite is discharged at the bottom of a drilled shot hole. The time required for shock waves reflected by subsurface rock to reach geophones placed at spaced distances at the surface is measured (Figure 2-6). The time recorded at the various locations can be used to determine where there are formations that may be water-bearing. The resistivity test measures the ground's electrical resistance. In general, the lower the electrical resistance, the greater the probability that water is present. Both of these tests are conducted on the ground surface, without the need to drill test holes.

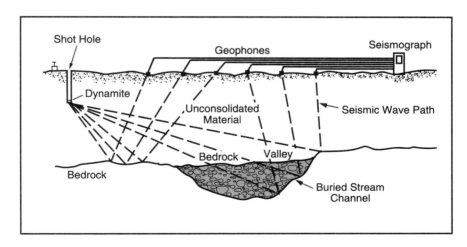

FIGURE 2-6
Application of the seismic refraction method for reconnaissance mapping

Test Wells

If it appears likely that water is present, exploration holes are drilled in key locations, and samples are kept of the material removed as drilling progresses on each hole. The holes are then logged electrically or by using gamma rays. Electric logging measures the change in the earth's resistance to an electrical current as the depth increases. Gamma-ray logging measures how controlled radiation penetrates the earth. Information from the logs is correlated with the samples removed during the drilling to identify the various formations below the surface.

Computer Modeling

Hydrologists interested in evaluating the many complex stresses and effects on an aquifer system can now use computer programs to perform studies that were not previously possible with manual calculations (Figure 2-7). Computer analyses using available information on aquifer recharge and withdrawals can be used to rapidly forecast the location of additional wells, the effects of contamination sources on the aquifer, and many other factors. At present, computer models are most widely used by government agencies to develop area and regional plans for groundwater use.

FIGURE 2-7
A computer being used to model groundwater conditions

Sources of Groundwater Contamination

New wells must be located a safe distance from sources of potential contamination. Existing wells sometimes become contaminated because of new contamination sources or sources that were not anticipated when the well was installed. Potential sources of groundwater contamination are listed in Table 2-1.

These contaminants may be naturally occurring or synthetic and may be located on or below the ground surface. If the contaminants were injected into the soil, it could have been done through ignorance, by accident, or by

TABLE 2-1   Potential sources of groundwater contamination

Source	Possible Major Contaminants
Landfills	
Municipal	Heavy metals, chloride, sodium, calcium
Industrial	Wide variety of organic and inorganic constituents
Hazardous-waste disposal sites	Wide variety of inorganic constituents (particularly heavy metals such as hexavalent chromium) and organic compounds (pesticides, solvents, polychlorinated biphenyls)
Liquid waste storage ponds (lagoons, leaching ponds, and evaporation basins)	Heavy metals, solvents, and brines
Septic tanks and leach fields	Organic compounds (solvents), nitrates, sulfates, sodium, and microbiological contaminants
Deep-well waste injection	Variety of organic and inorganic compounds
Agricultural activities	Nitrates, herbicides, and pesticides
Land application of wastewater and sludges	Heavy metals, organic compounds, inorganic compounds, and microbiological contaminants
Infiltration of urban runoff	Inorganic compounds, heavy metals, and petroleum products
Deicing activities (control of snow and ice on roads)	Chlorides, sodium, and calcium
Radioactive wastes	Radioactivity from strontium, tritium, and other radionuclides
Improperly abandoned wells and exploration holes	Variety of organic, inorganic, and microbiological contaminants from surface runoff and other contaminated aquifers

Source: Basics of Well Construction Operator's Guide.

intentional illegal disposal. Contaminants can be carried into an aquifer along with percolating surface water and can travel considerable distances under certain conditions.

Problems caused by contaminants can include disagreeable tastes and odors, presence of disease-causing organisms, and levels of chemical contaminants above recommended health-related limits for drinking water.

Because of differences in soil permeability, the depth of soil, and types of underlying materials, as well as the varying rates and directions of groundwater movement, there is no set safe distance that applies to all wells. Some chemical contaminants have been tracked more than 0.5 mi (0.8 km) from their source.

When siting a new well, one must keep in mind that when the well is put into operation, the direction of groundwater flow could change. As illustrated in Figure 2-8, water flow in an aquifer that appears to be coming from a direction with no contamination can actually be reversed by a new well's zone of influence.

Site Selection

Eventually, all available data are combined for study and the test site is selected. Unfortunately, the best site from a geological standpoint may not always be available for installation of a well. In a residential community, for instance, the decision of where to site a new well must also be based on where land can be acquired and where the construction of a well will be aesthetically acceptable to residents.

Types of Wells

Wells are holes sunk into the earth to obtain water from an aquifer. They are generally classified by the type of construction as follows:

- dug wells
- bored wells
- driven wells
- jetted wells
- drilled wells

Dug Wells

A dug well can furnish large quantities of water from shallow groundwater sources. Small-diameter wells are generally constructed manually with pick and shovel and larger wells with machinery such as a clamshell bucket

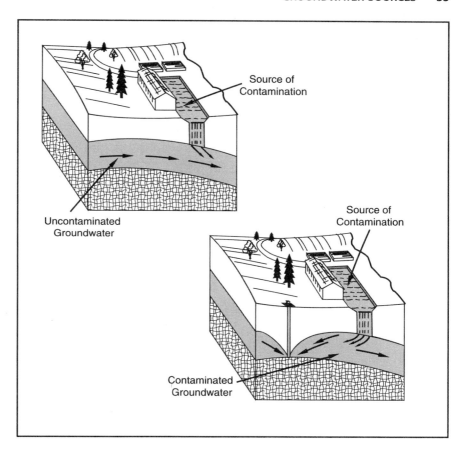

Source of
Contamination

Uncontaminated
Groundwater

Source of
Contamination

Contaminated
Groundwater

FIGURE 2-8
Reversal of flow
in an aquifer due
to well drawdown

if suitable soil conditions are encountered. Figure 2-9 illustrates a typical dug-well installation.

    If the exposed soil will stand without support, it may not be necessary to line the excavation until the water table is reached. Precast or cast-in-place concrete liners, commonly called curbs, are used to seal the shaft from groundwater contamination. The liners in contact with the water-bearing layers are perforated to allow water to enter. If the aquifer is sandy material, a layer of gravel is placed around the curb to act as a sand barrier.

    Yield from a dug well increases with an increase in diameter, but the increase is not directly proportional. Dug wells serving a public water system may be 8 to 30 ft (2 to 9 m) in diameter and from 20 to 40 ft (6 to 12 m) deep.

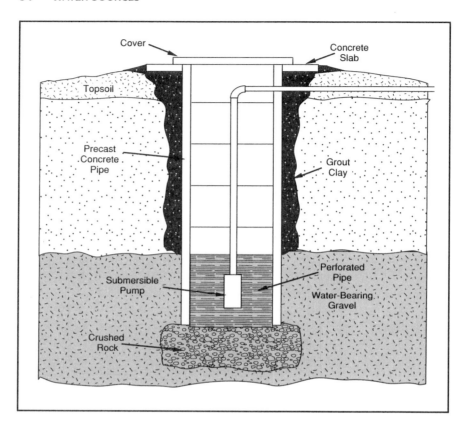

FIGURE 2-9
Construction of a dug well

Most dug wells do not penetrate much below the water table because of the difficulty of excavating in the saturated soil. For this reason, a dug well may fail if the water level recedes during times of drought or if there is unusually heavy pumpage from the well.

The opening of a dug well has a large surface area, making protection from surface contamination difficult. For that reason, state regulatory agencies usually classify these wells as being vulnerable to contamination, and they must be treated as a surface water source. Disinfection and possibly filtration treatment will be required if the water source is to be used by a public water system.

Bored Wells

Wells can be constructed quickly by boring if the soil types are suitable. The soil must be soft enough for an auger to penetrate, yet firm enough so it will not cave in before a liner can be installed. The most suitable formations for bored wells are glacial till and alluvial valley deposits.

The well is constructed by driving an auger into the earth. Bored wells are limited to approximately 3 ft (1 m) in diameter and depths of 25 to 60 ft (8 to 18 m) under suitable conditions. As the auger penetrates, extensions are added to the drive shaft. A casing is forced into the hole as material is removed until the water-bearing strata is reached. The well is completed by installing well screens or a perforated casing in the water-bearing sand and gravel.

Cement grout is used to surround the casing to prevent entrance of surface water, which could cause contamination. Bored wells are not frequently used for public water supplies.

Driven Wells

Driven wells are simple to install, but they are practical only when the water-bearing soils are relatively close to the surface and no rock layer or boulders exist in the soil between the surface and the aquifer. These wells consist of a pointed well screen, called a drive point, and lengths of pipe attached to the point. The point has a steel tip that enables it to be pounded through pebbles or disintegrated rock (otherwise known as hardpan) to the water-bearing soil (Figure 2-10).

The diameter of the well pipe varies from as small as $1\frac{1}{4}$ in. (32 mm) up to 4 in. (107 mm). The maximum depth that can be achieved is generally 30 to 40 ft (9 to 12 m). Because of the relatively small diameter of the pipe, water production from a single well will not usually be sufficient for the needs of a public water system. However, a battery of points may be used for greater production, with several wells connected by a common header to a suction-type pump. A suction pump can be used only when the static water level is no deeper than about 15 ft (5 m).

Construction of a driven well is usually begun by driving a pilot hole or outer casing. This casing of durable pipe can be pounded into the earth or placed in a bored hole. The casing prevents water contamination if there should be leaky joints in the well pipe.

After the outer casing is in place, the inner casing with a perforated well point is inserted. The drive point is then driven into the water-bearing formation.

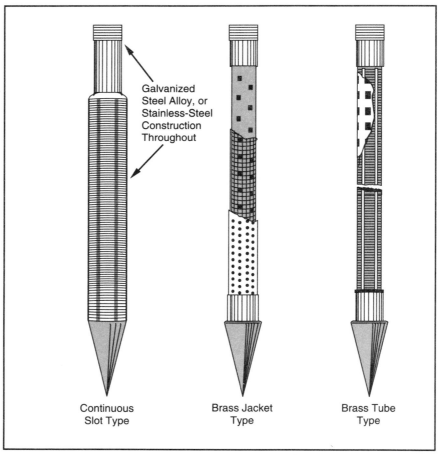

Galvanized
Steel Alloy, or
Stainless-Steel
Construction
Throughout

Continuous
Slot Type

Brass Jacket
Type

Brass Tube
Type

FIGURE 2-10
Different types of
driven-well points

Source: Manual of Individual Water Supply Systems *(1982).*

Jetted Wells

Wells generally cannot be constructed by jetting through clay or hardpan or where there are boulders in the soil. The equipment consists of a jetting pipe, which is equipped with a cutting knife on the bottom. Water is pumped down the pipe and out of the drill bit against the bottom of the hole.

The casing is usually sunk as the drilling progresses, until it passes through the water-bearing formation. The well screen connected to a smaller-diameter pipe is then lowered into the casing, and the outer casing is withdrawn to expose the screen to the formation.

Drilled Wells

Drilled wells are the type most commonly used for public water supplies because of the extreme depths and well diameters (up to 4 ft [1.5 m] or possibly larger) that can be required.

A drilled well is constructed using a drilling rig that makes the hole and a casing that is placed in the hole to prevent the walls from collapsing. Screens are installed when water-bearing formations are encountered, at one or more levels. The following are the more commonly used methods of drilling water supply wells:

- cable tool method
- rotary hydraulic method
- reverse-circulation rotary method
- California method
- rotary air method
- down-the-hole hammer method

Cable Tool Method

The percussion drilling method, commonly referred to as the cable tool method, is used extensively for wells of all sizes and depths. There are many commercial varieties of cable tool rigs. The operating principle for any variety is that there is a bit at the end of a cable that is repeatedly raised and dropped to fracture the soil material. The drilling tool has a clublike chisel edge that breaks the formation into small fragments. The reciprocating motion of the drilling tool mixes the loosened material into a sludge.

In each run of the drill, a depth of 3 to 6 ft (1 to 2 m) of the hole is drilled. The drill is then pulled from the hole, and a bailer is used to remove the sludge. The bailer consists of a 10- to 25-ft (3- to 8-m) long section of casing that is slightly smaller in diameter than the drilled hole and has a check valve in the bottom. A casing is forced into the hole as soon as it is necessary to prevent a cave-in of the walls.

As the drill operates, an operator continually adjusts the length of stroke and rapidity of blows based on experience and on the feel of the vibrations. The operator can also generally distinguish the hardness of the formation being drilled by the vibrations in the cable and can sense the passing of the drill from one formation to another. Samples of the cuttings are also periodically taken to check on the type of formation being penetrated.

After drilling the well to the maximum desired depth, a screen is lowered inside the casing and held in place while the casing is pulled back to expose the screen (Figure 2-11). The top of the screen is then sealed against the casing by expanding a packer of neoprene rubber or lead. When a well reaches consolidated rock, normal practice is to "seat" the casing firmly in the top of the rock and drill an open hole to the depth required to obtain the needed yield (Figure 2-12).

Rotary Hydraulic Method

In the rotary method of well drilling, the hole is made by rapid rotation of a bit on the bottom of multiple sections of drill pipe. The speed of rotation

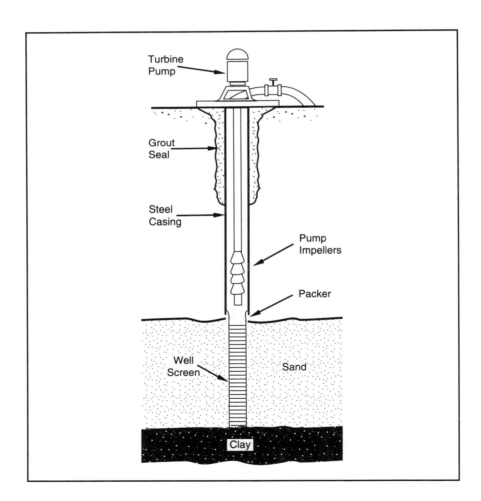

FIGURE 2-11
Exposed well
screen

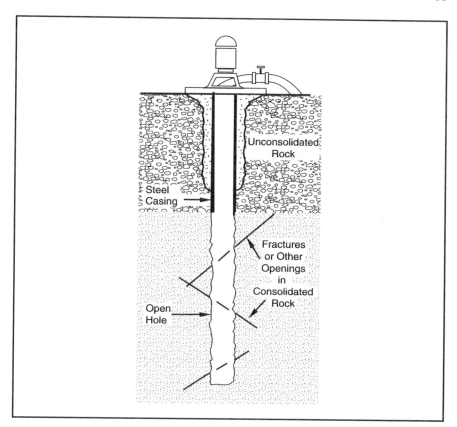

FIGURE 2-12
Casing seated at
top of rock layer
with an open hole
underneath

can be varied to achieve the best cutting effectiveness for different formations.

The drill pipe is hollow so that prepared fluid can be pumped down to the bit (Figure 2-13). The fluid flows out through holes in the bit, picks up loosened material, and carries it up the bore hole to the surface. The circulating fluid also helps to cool the bit. The fluid that flows to the surface overflows from the well and is routed by a ditch into a settling pit where the cuttings settle out. The fluid can then be reused.

Clay is usually added to the drilling fluid. It will adhere to the sides of the hole, and together with the pressure exerted by the drilling fluid, it prevents a cave-in of the formation material. In some cases, native clay can be used; otherwise, prepared materials such as bentonite or a special powdered material is used. The drilling fluid is prepared as a slurry in a pit near the

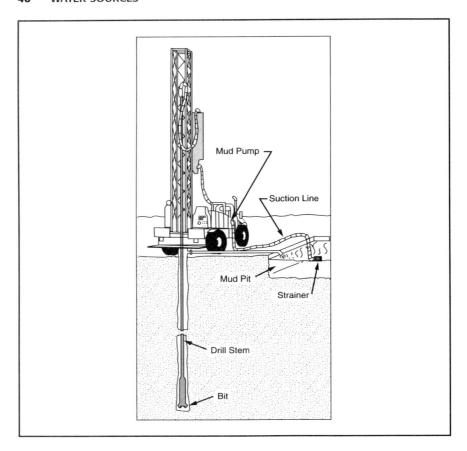

FIGURE 2-13
Circulation of
drilling fluid in a
direct rotary
drilling rig

drilling operation prior to being pumped into the well. When clay is being drilled through, the drilling fluid is generally just water.

Reverse-Circulation Rotary Method

The reverse-circulation rotary method of drilling differs from the regular rotary method in that the drilling fluid is circulated in the opposite direction. The fluid is forced down through the bore hole and returns to the surface in the hollow drill pipe, carrying the cuttings with it. A pump with a capacity of 500 gpm (1,900 L/min) or more is required to keep the fluid moving at high velocity. If an abundant freshwater supply is available, the discharge can be diverted to waste and not recycled. Otherwise, the cuttings are allowed to settle in a large pit, and the fluid is recirculated.

The head of the water in the casing generally prevents the walls from caving in without having to use mud additives. Therefore, this method is particularly well-suited for constructing gravel-pack wells, because the drilling fluid mixture used in the regular rotary method tends to plug the walls of the well.

Both rotary methods can drill holes up to 5 ft (1.5 m) in diameter, and the drilling is usually faster than cable tool drilling.

California Method

The stovepipe method, or California method, was developed primarily for sinking wells in unconsolidated material such as alternating strata of clay, sand, and gravel. The process is similar to the cable tool method except that a special bucket is used as both bit and bailer. Each time the bit is dropped, some of the cuttings are trapped in the bailer. When the bailer is filled, it is raised to the surface and emptied.

The process also uses short lengths of sheet metal for casing. The casing is forced down by hydraulic jacks or driven by means of cable tools. After the casing is in place, it is perforated in place using special tools.

Rotary Air Method

The rotary air method of drilling is similar to the rotary hydraulic method except that the drilling fluid is air rather than a mixture of water and clay. The method is suitable only for drilling in consolidated rock. Most large drill rigs are equipped for both air and hydraulic drilling so that the method may be changed as varying strata are encountered.

Down-the-Hole Hammer Method

A method frequently chosen to drill wells into rock uses a pneumatic hammer unit that is attached to the end of the drill pipe. The hammer is operated by compressed air. The air also cleans the cuttings away from the bit and carries them to the surface. For most types of rock, this is the fastest drilling method available.

The drilling rig for this method must be furnished with a very large air compressor. Most standard rigs use 750 to 1,050 ft^3/min (0.35 to 0.49 m^3/s) of air at a pressure of 250 psi (1,700 kPa). Some rigs are also capable of operating at 350 psi (2,400 kPa), which will advance the drill twice as fast as a standard rig. A drilling rig of this type is illustrated in Figure 2-14.

FIGURE 2-14
A modern
well-drilling rig

Courtesy of Ingersoll-Rand Company

Special Types of Wells

Unlike the previously described wells, there are several other types that are not named according to the method of construction. These types include

- radial wells
- gravel-wall wells
- bedrock wells

Radial Wells

Radial wells are commonly used near the shore of a lake or near a river to obtain a large amount of relatively good-quality water from adjacent sand or gravel beds. Radial wells are also used in place of multiple vertical wells to obtain water from a relatively shallow aquifer. A radial well can be described as a dug well that has horizontal wells projecting outward from the bottom of the vertical central well (Figure 2-15). The central well, or caisson, serves as the water collector for the water produced by the horizontal screened wells.

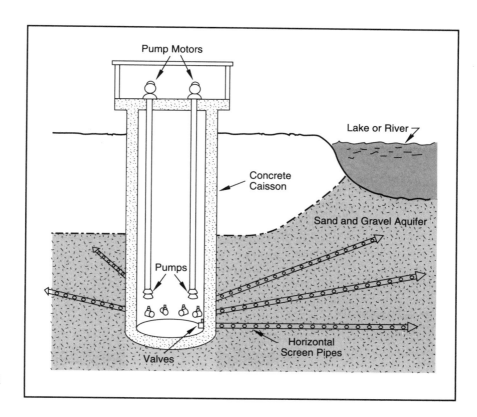

**FIGURE 2-15
Details of a radial well**

Construction begins by sinking the central caisson, which is generally 15 to 20 ft (5 to 6 m) in diameter. The caisson is made by stacking poured-in-place reinforced concrete rings, each about 8 to 10 ft (2 to 3 m) high. The first section is formed with a cutting edge to facilitate the caisson's settling within the excavation. As soil is excavated from within the caisson, the concrete ring sinks into place. Additional sections are then added and the excavation progresses. When the desired depth is reached, a concrete plug is poured to make a floor.

Horizontal wells are then constructed through wall sleeves near the bottom of the water-bearing strata. The laterals may be constructed of slotted or perforated pipe or they may have conventional well screens. Each horizontal well is constructed with a gate valve located inside the caisson for subsequent dewatering or to discontinue use of individual collector wells. A superstructure is then erected on top of the caisson to house pumps, piping, and controls.

Gravel-Wall Wells

Gravel-wall wells are also commonly called gravel-packed wells. They are best used in formations composed of fine material having uniform grain size. As illustrated in Figure 2-16, a bed of gravel is installed around the screen, which in effect gives greater surface area for the infiltration of water into the well, while effectively blocking the entrance of sand.

The most common construction method is to install a large-diameter casing into the water-bearing strata and then lower a small casing with a well screen into the hole. The area around the screen is then filled with gravel, and the outer casing is withdrawn corresponding to the length of the screen. The gravel used must be clean, washed, and composed of well-rounded particles that are four to five times larger than the median size of the surrounding natural material. The size and gradation of the gravel are critical in effectively blocking the entrance of fine sand.

Other methods of constructing gravel-wall wells include underreaming, bail-down, and pilot-hole. In the underreaming method, a well is drilled to the top of the water-bearing formation and the casing set. The formation is then enlarged to a greater diameter using special tools, and the screen and gravel are put in place. The bail-down method of construction uses a cone-bottomed screen that is bailed into place while gravel is fed into the space between the inner and outer casing. In the pilot-hole method, the gravel is fed into the formation through small pilot wells evenly spaced around a central well as the fine sand is withdrawn through the central well.

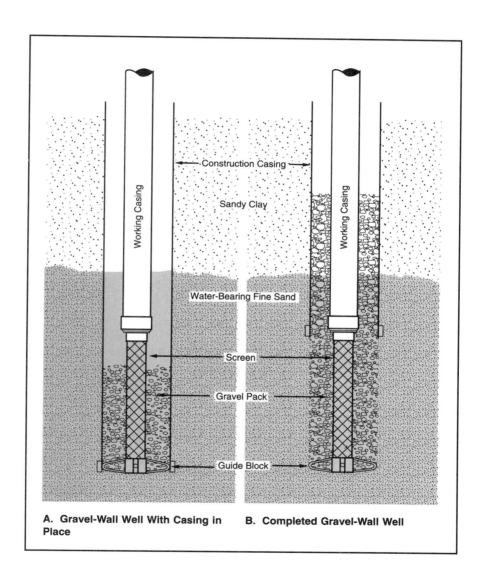

FIGURE 2-16
Gravel-wall well
construction

A. **Gravel-Wall Well With Casing in**
Place

B. **Completed Gravel-Wall Well**

Bedrock Wells

Bedrock wells are drilled into the underlying bedrock. Water flows to the well through fractures in the bedrock rather than from a saturated layer.

Well Construction Procedures

Wells can be constructed using a number of different procedures. The principal differences in the method used are how deep the well must be and whether the material that must be penetrated is gravel, clay, or rock. After a well has been constructed, it is then usually necessary to specially treat the well to obtain optimum productivity and water quality. The last step in well construction is to run pumping tests to confirm the well's capacity.

Well Components

The components that are common to most wells are

- well casings
- well screens
- grouting

Well Casings

The well casing is a lining for the drilled hole that maintains the stability of the open hole from the land surface to the water-bearing formation. A secondary function of a casing is to prevent surface water from contaminating the water being drawn from the well.

Materials commonly used for well casings include wrought iron, alloyed or unalloyed steel, ingot iron, fiberglass, and plastic. The principal factors in determining the suitability of a casing material are the stress that will be placed on the casing during the installation and the corrosiveness of the water and soil that will be in contact with the casing.

Many weights of steel and wrought-iron casing material are available for use under varying conditions. Lightweight materials may be used for test wells and temporary wells.

When a well is being drilled by the cable tool method, the casing is driven as soon as it becomes necessary to prevent the walls of the well from caving in. A drive shoe of hardened steel is attached to the lower end of the pipe, and a drive head is attached to the top of the pipe to withstand the hard blows of driving the casing into the ground. As the drilling progresses, the casing is continually pounded by the action of the drilling equipment.

Wells constructed through rotary methods are not usually cased until the well hole has been completed. A casing diameter smaller than the hole is used so that the casing does not have to be driven.

If additional protection from corrosion and surface water pollution is required, an outer casing is first installed, an inner casing is lowered into place, and the space between them is filled with cement grout. The outer casing may be left in place, or it may be withdrawn as the grout is added.

Well Screens

Wells completed in unconsolidated formations such as sand and gravel are usually equipped with screens (Figure 2-17). A properly sized screen will allow the maximum amount of water from the aquifer to enter the well with a minimum of resistance, while preventing the passage of sand into the well.

The selection of the proper screen requires highly specialized knowledge and experience. Based on samples of material from the water-bearing formation, the screen can usually be sized to have openings larger than the finest surrounding material. During development of the well, the fines are removed while larger particles are held back, forming a graded natural-gravel barrier around the screen (Figure 2-18).

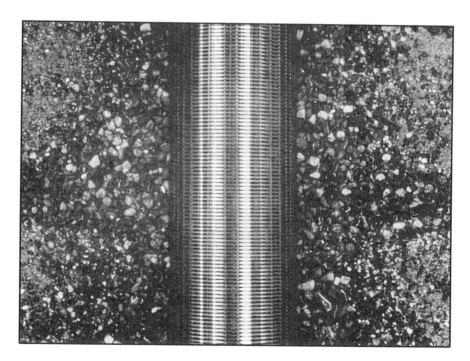

FIGURE 2-17
Properly designed
well screen

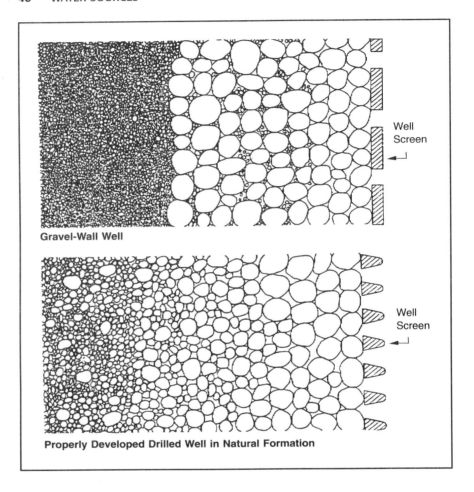

Well
Screen
←⌐

Gravel-Wall Well

Well
Screen
←⌐

Properly Developed Drilled Well in Natural Formation

FIGURE 2-18
Coarse material
used to hold fine
sand away from a
well screen

The size of a screen, or the slot number, is usually expressed in thousandths of an inch or millimeters. The slot size is usually selected to permit between 35 and 60 percent of the formation material to pass through it, depending on the uniformity of the grains.

It is important to keep the velocity of water entering the screen opening between 0.1 and 0.3 ft/s (0.03 and 0.09 m/s) (and not lower) to minimize head losses and chemical precipitation. The open area of the screen after the well has been developed must be carefully estimated because up to 50 percent of the screen slots may be plugged by formation particles. After the slot size has been determined, either the screen length or diameter must be adjusted to provide the total required screen opening.

Well screens are available in various materials, including plastic, mild steel, red brass, bronze, and stainless steel. The type of material used depends on the type of soil, corrosivity of the water, cleaning and redevelopment methods, and other factors. Information and specifications for screen selection are furnished in AWWA A100, *Standard for Water Wells* (latest edition).

Grouting

Wells are cemented, or grouted, for the following reasons:

- to seal the well from possible surface water pollution
- to seal out water from water-bearing strata that have unsatisfactory quality
- to protect the casing against exterior corrosion
- to restrain unstable soil and rock formations

When a well is drilled, an annular space is normally formed around the casing. This space must be sealed in order to prevent contaminants from entering, either from the land surface or from formations with crevices connected with the surface (Figure 2-19). In formations that may cave in, such as sand, the cavity will tend to seal itself eventually, but in stable formations, the cavity may remain indefinitely if not filled.

When corrosion of the casing is likely, a well may be purposely drilled larger than the casing. This forms a cavity around the casing that can be filled with about 2 in. (50 mm) of grout.

Portland-cement grout is usually mixed in a ratio of about one bag of cement to 5.5 gal (21 L) of clean water. Special additives are often used to accelerate or retard the time of setting and provide other special properties to the grout.

The grout must be placed in the cavity, starting at the bottom and progressing upward so that no gaps or voids exist in the seal. Depending on the size of the annular space, a dump-bailer, water-pressure driving, pumping, or a tremie may be used to place the grout. In some areas, bentonite clay can be used instead of cement to fill the voids around the casing. Grout must be placed in compliance with state regulations.

Well Development

After a well is constructed, various techniques are used to *develop* the well — that is, to allow the well to produce the best-quality water at the

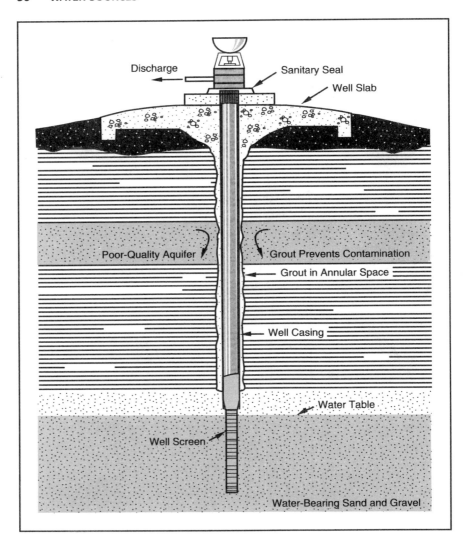

FIGURE 2-19
Sealed annular
space that
prevents
contamination

highest rate from the aquifer. The method used is dependent on many different factors and is usually selected by the well construction contractor. Methods commonly used include

- high-rate pumping
- surging
- increased-rate pumping and surging

- use of explosives
- high-velocity jetting
- chemical agents
- pressure acidizing
- hydraulic fracturing ("fracking")

High-Rate Pumping

When a well draws water from a rock aquifer where no screen is used, development usually involves simply pumping the well at a rate higher than the normal production rate. This is done to flush out the bore and surrounding aquifer.

Surging

Surging generally employs a plunger, or surge block, on a tool string, which is moved up and down in the well casing (Figure 2-20). This creates flow reversals in the well face, with the result that fine materials are dislodged and fall to the bottom of the well. The fines are then removed with a bailer. Various types of surge blocks are used.

Surging is also achieved by use of compressed air. A provision is made to close the top of the casing so that applied air pressure will drive the water in the well through the screen or exposed walls of the well. A jet of compressed air is then used to raise fine material that has been freed, up to the surface for removal. When this system is used, care must be taken not to drive air into the formation because it may collect in pockets that will reduce the aquifer production.

Increased-Rate Pumping and Surging

Increasing the pumpage in combination with surging is effective in developing wells in unconsolidated limestone. As much of the drilling fluid as possible is first removed from the well, and as the water clears, gentle surging assists in development. Pumping rates are gradually increased in steps to develop the well fully.

Use of Explosives

Explosives may be used in well development to fracture massive rock formations. Heavy charges are set off in the well bore at predetermined points to radiate fractures outward from the bore. Another technique is to shoot dynamite in the most permeable portions of a formation to clear drilling mud or other foreign materials that are plugging the bore face.

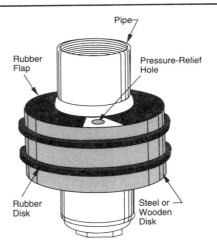

A. Details of a Typical Surge Block

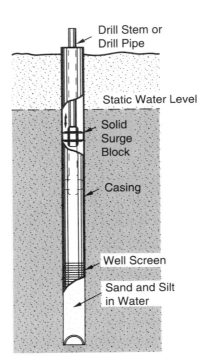

B. Surge Block Being Used to Develop a Well by Pulling Silt and Fine Sand Through the Screen

**FIGURE 2-20
Surge block
schematics**

In another method, very light charges are set off in a series of explosions to produce vibrations that agitate the materials surrounding the bore. With any of these methods, the loosened material must be removed from the well by bailing.

High-Velocity Jetting

Hydraulic jetting involves pumping clean water at high pressures against the sides of the well so that the nozzle velocity is about 150 ft/s (45 m/s). When this force strikes the formation, it can break down dense compacted materials and mud left from rotary drilling. The jets are rotated at intervals of from 6 to 12 in. (150 to 305 mm) throughout the water-entry area of the well. Pumping out of the well is simultaneous at a rate somewhat higher than the rate at which jet water is being introduced.

Chemical Agents

It has been found that chemical additives frequently enhance the effectiveness of various development methods. For instance, wetting agents used with hexametaphosphates are particularly useful in increasing the effectiveness of jetting. Acids may also be used where there are iron or carbonate deposits that must be dispersed.

Pressure Acidizing

In rock aquifers where there is good isolation from overlying formations, pressure acidizing may increase well productivity. High concentrations of acid, usually inhibited muriatic acid, are forced under high pressure into the formation.

Hydraulic Fracturing

Hydraulic fracturing is a process in which a fluid is forced into the well under pressure great enough to open the separations between strata and along existing fractures. Gelling agents and selected sand are added to the fluid to hold the fractures open after pressure is removed. This process can be used only in rock aquifers, and the well casing must be firmly sealed and anchored in place.

Pumping Tests

After a well is developed and the quality of water produced is satisfactory, pumping tests are performed. These practical tests are intended to confirm that the well will produce at its design capacity. Most water supply well contracts require a guaranteed yield and may also stipulate that a certain level of well efficiency be reached. Contracts also usually specify the duration of the drawdown test that must be conducted to demonstrate that the yield requirement is met.

Many well acceptance tests are conducted with a temporary pump installed, usually powered by a gasoline or diesel engine. The test period for public water supply wells is generally at least 24 hours for a confined aquifer and 72 hours for an unconfined aquifer.

A means of measuring the flow rate of the pumped water must be provided. A commercial water meter, an orifice or weir, or periodic checking of the time required to fill a container of known volume are all methods of measuring flow rate. The pumping rate must be held constant during the test period, and the depth to the water level in the well must periodically be measured. Drawdown is usually measured frequently during the first hour of the test, and is then measured at gradually longer intervals as the test period progresses.

Sanitary Considerations

Whenever possible, wells should be developed from formations deep enough to protect the water from surface contamination. Where a shallow groundwater source must be used, surface water contamination can be reduced if an impervious surface is created over the ground surface surrounding the well. One method is to place a 2-ft (0.6-m) layer of clay having a radius of about 50 ft (15 m) around the well. Contamination of wells by surface water flowing along the exterior of the casing is prevented by filling the space between the casing and the hole with cement grout. A concrete cap or platform should then be put in place around the well. The construction methods for wells are detailed by state regulatory agencies and, in some cases, by local jurisdiction as well. The requirements must be followed by the well driller, and the installation must be approved before the well is certified for use.

It is almost impossible to completely avoid contamination of the soil, tools, aquifer, casing, and screen during construction, so the water from the well, even after development, is likely to be contaminated. Extended pumping will usually rid the well of contamination, but disinfection with a chlorine solution is quicker and more desirable.

A well is disinfected by the addition of sufficient chlorine to give a concentration of 50 mg/L. The pump can then be started and stopped to surge the disinfectant into and out of the aquifer and through all the well and pump components. Gravel packing is difficult to disinfect by ordinary procedures, so powdered or tablet calcium hypochlorite is usually added when the gravel is being placed.

For more details and specific procedures, refer to AWWA C654, *Standard for Disinfection of Wells* (latest edition).

Aquifer Performance

To measure changes in an aquifer, small-diameter test wells called observation wells are installed. There is no set number of observation wells that should be used. One well positioned near an operating well will sometimes be sufficient, but several may be required under different circumstances.

Aquifer Evaluation

Observation wells are located carefully on a map. The distances measured between the well being tested and the observation wells, and between the observation wells themselves, are also recorded. The water level in each observation well is periodically measured relative to an elevation reference point, or benchmark.

One method of measuring the water depth in a test well is to use a steel tape graduated in tenths and hundredths of a foot (or metric units), with a weight attached (Figure 2-21A). The tape is chalked and lowered into the well until the weight reaches the bottom. When the tape is withdrawn, the water depth is indicated by the wetted position on the tape. Electronic depth-measuring devices are also available. They include a probe that is lowered into the well to activate a signal when water is contacted (Figure 2-21B).

Another way of measuring water depth, effectively used for many years for fixed installations, is the air-pressure tube method (Figure 2-21C). A small-diameter tube is suspended in the well beneath the water level, and the exact distance from the bottom of the tube to the ground surface is known. The top of the tube is fitted with a pressure gauge and a source of compressed air. Determining the water depth involves forcing air into the tube until it bubbles from the submerged end of the tube. The pressure gauge is then read, and the submerged length of tube can be calculated from the pressure required to displace the column of water in the tube. The calculation method is detailed in *Basic Science Concepts and Applications,* part of this series.

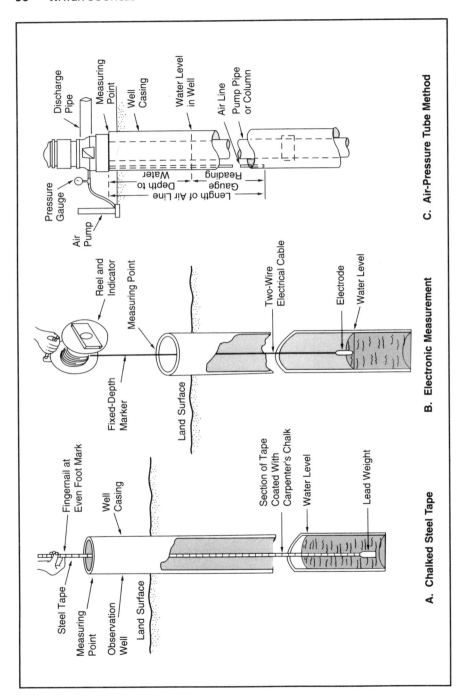

**FIGURE 2-21
Methods of
measuring water
level depth in
wells**

Evaluating Performance

Aquifer performance can be evaluated by the following three methods:

- drawdown method
- recovery method
- specific-capacity method

In the drawdown method, the production well is pumped and water levels are periodically observed in two or more observation wells. The data are plotted and can be analyzed by various methods to relate drawdown in feet (or meters) to time measured in hours or days at a specific pump rate.

The recovery method involves measuring the change in water level in an observation well after the pumping has been stopped.

The specific-capacity method involves a relatively short test. As discussed earlier, specific capacity is the well yield per unit of drawdown. It does not indicate aquifer performance as completely as the other tests. However, it is especially valuable in evaluating well production after a period of time and in making comparisons with the data from when the well was new.

Well Operation and Maintenance

When a well is in service, it is important to keep records that will indicate any changes that might predict future problems. The following tests should be conducted and data recorded for each well:

- static water level after the pump has been idle for a period of time
- pumping water level
- drawdown
- well production
- well yield
- time required for recovery after pumping
- specific capacity

Conditions for the above tests should be the same each month so that direct comparisons can be made.

Regular maintenance of all the structures, equipment, and accessories is important to provide long-term, trouble-free service. Wells can fail if the casing or screen collapses or corrodes through. Probably the most common well operational problem is plugging of the screen. The causes can be mechanical, chemical, or bacteriological in nature. It is usually best to get professional assistance in correcting the problem. If the screen is accidentally damaged, repair can be extremely expensive, or the entire well may have to be replaced.

Bacteriological samples should periodically be collected directly from each well and tested. If there are indications of contamination, periodic disinfection of a well may be necessary to prevent growth of nuisance bacteria that can lead to production problems.

Well Abandonment

Test holes must be abandoned when they are no longer of use, and occasionally a production well must be abandoned. This must be done properly for the following reasons:

- to eliminate a physical hazard
- to prevent groundwater contamination
- to conserve the aquifer
- to prevent mixing of desirable and undesirable water between aquifers

The basic concept governing the proper sealing of abandoned wells is the restoration of the geologic and hydrologic conditions that existed before the well was constructed. Each well that is abandoned should be considered unique. The methods used should be those that will give the best results. Recovery of pumps, casings, screens, and other hardware is best performed by the firm that installed the well. The well construction company is also likely to be the best firm to complete the abandonment procedures properly.

Selected Supplementary Readings

AWWA Standard for Disinfection of Wells, **ANSI/AWWA C654.** Denver, Colo.: American Water Works Association.

AWWA Standard for Water Wells, **ANSI/AWWA A100.** Denver, Colo.: American Water Works Association.

Borch, M.A., S.A. Smith, and L.N. Noble. 1993. *Evaluation and Restoration of Water Supply Wells.* Denver, Colo.: American Water Works Association Research Foundation and American Water Works Association.

Driscoll, F.G. 1986. *Groundwater and Wells.* St. Paul, Minn.: Johnson Filtration Systems Inc.

Manual M21, Groundwater. 1989. Denver, Colo.: American Water Works Association.

Manual of Individual Water Supply Systems. 1982. US Environmental Protection Agency, Office of Drinking Water. EPA-570/9-82-004. Washington, D.C.: US Government Printing Office.

Manual of Instruction for Water Treatment Plant Operators. 1975. Albany, N.Y.: New York State Department of Health.

Manual of Water Utility Operations. 8th ed. 1988. Austin, Texas: Texas Water Utilities Association.

Water Quality and Treatment. 4th ed. 1990. New York: McGraw-Hill and American Water Works Association (available from AWWA).

CHAPTER 3

Surface Water Sources

Although groundwater is used as a water supply source in rural areas and by most small communities, it is rare for sufficient groundwater to be available to serve large cities. Consequently, almost all large population centers must be served from surface water sources. Many large cities have thrived because of the availability of fresh water. Examples include the large cities around the Great Lakes and along major rivers. When a city continues to grow in spite of insufficient groundwater or surface water available to supply public and industrial needs, it is necessary to reroute surface water to the city from other areas through canals or pipelines.

Surface Runoff

Most surface water originates directly from precipitation in the form of rainfall or snow. Rainfall runs rather quickly off the land surface toward streams. Snow, on the other hand, is a form of water storage that is of extreme importance. Most areas of the western United States and Canada depend on water melting from the snowpack in the mountains for a continuing supply of water during the warm months.

Groundwater from springs and seeps also contributes flow to most streams. If the adjacent water table is at a higher level than the water surface of the stream, water from the water table will flow to the stream. Many streams would dry up shortly after a rain if it were not for groundwater flow.

Influences on Runoff

The principal factors that affect how rapidly surface water runs off the land are

- rainfall intensity
- rainfall duration
- soil composition
- soil moisture
- ground slope
- vegetation cover
- human influences

Rainfall Intensity

Slow, gentle rainfall generally produces very little runoff. There is plenty of time for the rain to soak into (infiltrate) the soil. However, as rainfall intensity increases, the surface of the soil becomes saturated. Because a saturated soil can hold no more water, further rainfall builds up on the surface and begins to flow, creating surface runoff.

Rainfall Duration

The duration of a rainstorm also influences the amount of runoff. Even a slow, gentle rain, if it lasts long enough, can eventually saturate the soil and allow runoff to take place.

Soil Composition

The composition of the surface soil also has a marked affect on the amount of runoff produced. Coarse sand, for example, has large void spaces that allow water to pass through readily, so even a high-intensity rainfall on sand may result in little runoff. On the other hand, light rainfall on soil such as clay will generally result in a relatively large amount of runoff. Clay soils not only have small void spaces but they swell when wet. This expansion closes the void spaces, which further reduces the infiltration rate and results in greater surface runoff.

Soil Moisture

If soil is already wet from a previous rain, surface runoff will occur sooner than if the soil were dry. Consequently, the amount of existing soil moisture has an effect on surface runoff. If the ground is frozen, it is essentially impervious, and runoff from rain or melting snow may then be up to 100 percent.

Ground Slope

Ground slope has a great effect on the rate of surface runoff. Flow of water off flat land is often so slow that there is usually ample opportunity for a large portion of it to soak into the ground. But when rain falls on steeply sloping ground, as much as 80 percent of it may become surface runoff.

Vegetation Cover

Vegetation covering the ground plays an important role in limiting runoff. Roots of plants and deposits of decaying natural organic litter (such as leaves, branches, or pine needles) create a porous layer above the soil and allow water to move easily into the soil. Hard, driving rains will sometimes compact bare soil, closing the void spaces. However, vegetation and organic litter act as a cover to protect soil from this compaction; they maintain the soil's water-holding and infiltration capacity. This cover also reduces evaporation of soil moisture, though this effect is less significant.

Human Influences

Human activities have a decided effect on surface water runoff. Although dams are generally built to retard the flow of runoff, most other activities and structures tend to increase the rate of flow. Canals and ditches are usually constructed to hasten flow, and agricultural activities generally remove vegetation that would naturally retard the runoff rate.

The impervious surfaces of buildings, streets, and other paved areas greatly increase the amount of runoff. By hastening the flow of surface water, flooding often occurs and the opportunity for water to percolate into the soil to replenish groundwater supplies is reduced.

Watercourses

Surface water runoff naturally flows along the path of least resistance. All of the water within a watershed flows toward one primary watercourse, unless it is diverted by a constructed watercourse such as a canal or pipeline.

Natural Watercourses

Typical natural watercourses include brooks, creeks, streams, and rivers. Watercourses may flow continuously, occasionally, or intermittently, depending on the frequency of rainfall and other sources of water.

Perennial streams. Watercourses that flow continuously at all times of the year are called perennial streams. They are generally supplied by a combination of surface runoff, springs, groundwater seepage, and possibly snowmelt.

Ephemeral streams. Streams flowing only occasionally are called ephemeral streams (Figure 3-1). The word *ephemeral* refers to something that has only a brief existence. These streams usually flow only during and shortly after a rain and are supplied only by surface runoff.

Intermittent streams. Many streams fall between the two other categories. Depending on the rainfall and snowmelt, they may have running water for weeks or months at a time. These streams, which are usually dry at some times, are known as intermittent streams.

Constructed Watercourses

Many facilities are constructed to hasten the flow of surface water or to divert it in a direction other than the one in which it would flow under natural conditions. Ditches, channels, canals, aqueducts, conduits, tunnels, and pipelines are all examples. In many cases, water is diverted from its natural watershed into another.

FIGURE 3-1 Bed of an ephemeral stream

Many local waterways are constructed to prevent water from ponding in areas where it would impede agricultural or urban development. Other examples of constructed watercourses include canals provided for boat access or shipping, aqueducts to provide irrigation water to arid regions, and pipelines to convey potable water to areas that have an insufficient natural supply.

Water Bodies

A water body is a water storage basin. It can be a natural basin such as a pond or lake, or a reservoir constructed to serve some specific water need. Reservoirs range from small basins constructed by farmers for watering livestock to massive reservoirs that store water for public water system and recreational use.

Surface Source Considerations

The development of a surface water source for use by a public water supply requires careful study of the quantity and quality of water available.

Quantity of Water Available

The prime consideration in selecting a water source is that the supply must reliably furnish the quantity of water required. The flow in any watercourse fluctuates in relation to the amount of rainfall and runoff that occurs, so it must be known that sufficient water will be available during low-flow or drought conditions.

Community water needs are normally greatest during the warm months of summer and early fall, and this is usually the period when natural stream flow is the lowest. To compensate for this problem and increase water supply reliability, impounding reservoirs can be constructed to store water during periods of excess supply. This water can then be used during periods when natural stream flow is deficient.

The *safe yield* is a concept that describes the availability of water based on the size of the impoundment, the area of the watershed, and the expected rainfall. In simple terms, the safe yield of a watershed area represents the maximum rate at which water can be withdrawn continuously over a long period of time. For example, without storage, the safe yield of a stream is the amount of water that can be withdrawn during a period of lowest flow.

Safe yield can be increased by the addition of storage. If a stream has a high enough flow in the winter or other season to replenish water in an impoundment, the amount of water that can be withdrawn during dry seasons can be greatly increased.

There are many interrelated factors that must be considered when the safe yield of a water source for public supply is being calculated. In addition to the rainfall and flow considerations, allowances must also be made for other uses of the source water for agricultural irrigation or by other water systems. Anticipated increases in usage due to population and business growth in coming years must also be considered.

Water Quality

Although it is technically possible to treat just about any poor-quality water to an acceptable quality for public water system use, it is often not economically practical. For example, it may be more cost effective to pipe water a considerable distance from a remote, good-quality supply than to treat poor-quality water that is available locally.

The economic concerns are often complex and are greatly influenced by supply and demand. There may also be political and legal considerations, such as who has the "rights" to the water. One good example is salt water, which is available to many large cities located adjacent to oceans. At this time, it is less expensive to transport fresh water from sources hundreds of miles away than to desalt ocean water. However, in areas of the world where fresh water is not sufficiently available to supply demand, desalinization is used.

The economics are changing as treatment technology improves. Improvements in membrane technology, for instance, will probably substantially reduce the cost of treating water from poor-quality sources in coming years.

Some of the principal quality factors that must be considered in evaluating the suitability of a water source for use are

- water temperature
- taste, odor, and color
- occurrence of algae growth
- excessive turbidity
- susceptibility to microbiological contamination
- susceptibility to chemical or radiological contamination

The influences that these factors have on water use and treatability are covered in *Water Treatment* and *Water Quality*, other titles in this series.

Water Storage

Water can be stored by natural means, in impoundments, or through recharging of natural aquifers.

Natural Storage

Natural lakes are often used as a source for public water supply, but a lake must usually be relatively isolated from human activities to have good water quality. Lakes having a watershed with heavy agricultural activity, industrial operations, or urban areas are often highly polluted.

If a lake located some distance from a community is being considered for use, it must be decided whether it is better to treat the water at the source and then pipe the finished water to the community or to pipe the raw water to the community and treat it there. Because of the considerable cost of piping water over long distances, an engineering analysis of all the alternatives must be made.

Impoundments

Until the mid-1800s, storage impoundments were constructed by digging out a shallow basin and using the excavated material to build up embankments (Figure 3-2). An alternative was to construct low dams across streams to create reservoirs.

With the development of modern construction equipment and techniques, dams have become a common way of creating a large reservoir. They are usually built where the reservoir can be formed by damming a valley. The

FIGURE 3-2
Excavated
impoundment

deeper and narrower the valley is, the easier it is to construct the dam. Dams are generally constructed either of earth and rock or of concrete.

Embankment Dams

Dams constructed of earth and rock are called embankment dams. They must be carefully engineered to resist the forces of the reservoir water and prevent damage or failure due to seepage or undermining.

A cross section of a typical earth dam is shown in Figure 3-3. An embankment dam may be constructed mostly of rock if locally available, but there must be an inner core of impermeable material to ensure water tightness.

Masonry Dams

Masonry dams are constructed of concrete. They are generally used where the required dam does not need to be very wide but must be quite high. Figure 3-4 illustrates a large concrete dam.

Dam Design

The subsurface conditions are among the most important considerations in dam construction. The soil below the dam must have sufficient strength and be impermeable enough that the dam will not fail because of undermining. Subsurface conditions are investigated using core borings. If necessary, unsafe material must be removed and replaced with suitable material before the dam is constructed.

Some small reservoirs have sufficient capacity to hold all water that flows into them, so no outlet is necessary. Most dams, though, are expected to overflow at least occasionally, and some release enough water at all times to maintain flows downstream of the dam.

**FIGURE 3-3
Cross section of
dam and intake
tower for
impounded
surface water
supply**

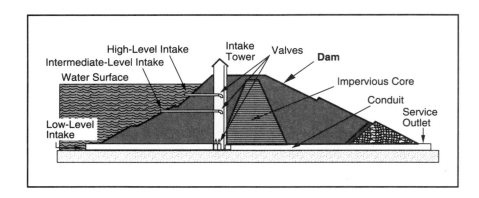

FIGURE 3-4
Chute spillway on
a concrete dam

The design of the dam outlet, or spillway, is particularly important in regulating the flow from a reservoir. Much of the time, the flow must be regulated periodically to maintain a balance between stream flow and the proper amount of water in storage. Small dams are provided with a simple slide gate that is raised and lowered to adjust flow. The spillway must also allow safe release of unusually heavy flows from the reservoir. If water should overflow the top of a dam, it will usually cause a washout that will severely damage or destroy the dam.

Dam Maintenance

Most very large dams have been constructed with federal funds and remain under federal ownership. Others are owned by the state or local jurisdiction or solely by a water utility.

In cases where a dam is owned by a water utility, the system operator may have responsibility for dam maintenance. The structure must periodically be inspected and maintained to prevent failure. Failure of a dam can result in both loss of a water source and destruction of downstream property. Water-controlling structures such as spillways and slide gates require periodic maintenance to keep them in good working order and to prevent failure.

Careful records should be kept of all inspections and work done on dams and water-control structures. Some states have specific procedures and reports that must be followed and maintained on dam maintenance.

Decreasing numbers of dams have been built in recent years, primarily because of public concern with regard to the environmental damage caused by flooding areas of scenic beauty and disrupting wildlife habitats.

Groundwater Recharge

Some attempts have been made to provide water storage by recharging depleted groundwater aquifers. Methods to accomplish this include providing large basins to allow the water to infiltrate the soil and the direct injection of water.

If untreated surface water is to be injected into the soil, care must be taken not to pollute the aquifer. In addition, turbidity in the injected water will gradually plug the injection area. Therefore, injection water must generally be fully treated to drinking water quality before it can be injected. Another problem with using recharge as a means of storing water is that the volume of water pumped into the aquifer cannot always be fully recovered.

In recent years, there has been increasing objection to the construction of new surface reservoirs for a variety of environmental reasons. In some cases, projects have been approved by local regulatory agencies, but the plans have been vetoed by the US Environmental Protection Agency. When storage of large quantities of water is necessary but the creation of new surface impoundments is completely unacceptable, the options will be either to pipe water for long distances to where a reservoir is acceptable or to store water in underground aquifers. As a result, aquifer recharge may become more common in the future.

All injection of fluids into the ground is now subject to state approval under the rules of the federal Underground Injection Control Program.

Intake Structures

Water is removed from a surface impoundment or stream by an intake structure. Types of intake structures include simple surface inlets, submerged intakes, movable intakes, pump intakes, and infiltration galleries.

Surface Intakes

In the simplest form, a surface intake consists of a small concrete structure containing a slide gate (Figure 3-5). When the slide gate is removed, water spills into a canal, ditch, or pipeline, which then carries the water to the

FIGURE 3-5
Surface diversion
with a slide gate

treatment plant. Larger installations may consist of a structure that has a bar screen and methods of removing accumulated trash from the screen.

The principal disadvantage of using a surface intake on a river or reservoir is that water quality is not usually as good as it would be below the surface. The water may be warmer in summer; have ice on the surface in winter; and may have algae, leaves, logs, and other floating debris at the surface at certain times of the year. If the elevation of the water in the lake or stream fluctuates, there is also the problem of locating a surface intake so that it will draw water under all conditions.

Many surface water treatment plants that normally draw water from a submerged intake also have an emergency surface, or "shore," intake. In the event of plugging or other failure of their submerged intake, the surface intake can be activated to furnish water of usable quality while repairs are made.

Submerged Intakes

Submerged intakes draw water from below the surface. The best water quality in most lakes and streams can usually be obtained in deep water. Intakes should not be located directly on the bottom or they will draw in silt and sand; therefore, they are often raised a short distance above the bottom. Submerged intakes have the advantages of presenting no surface obstruction to navigation and being less likely to be damaged by floating debris and ice. There are many designs used for submerged intakes; one typical type is illustrated in Figure 3-6.

Movable Intakes

Water from shallow depths may be quite warm in summer and may also have heavy algae growths. At these times, water drawn from deeper levels is more likely to be free of these problems.

On the other hand, deep water in lakes and reservoirs can sometimes develop poor quality in winter as a result of stratification, which occurs when colder and more dense water sinks to the bottom. Then, oxygen depletion and the dissolving of nuisance chemicals from bottom sediment can produce serious taste-and-odor problems. However, water drawn from higher levels is more likely to be free of these problems.

Movable intakes are used in lakes and streams where the water levels vary greatly or where there are other reasons for varying the depth of withdrawing water. Movable intakes are also used where good foundation is lacking or where other conditions prevent the construction of a more substantial structure.

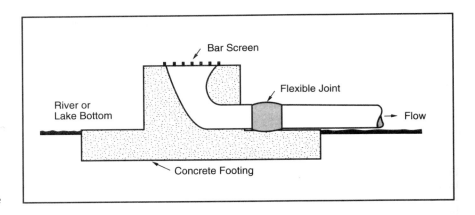

FIGURE 3-6
Schematic of a
submerged intake

Intake structures that have multiple inlet ports are also used in locations where the water level varies (Figure 3-7). By opening and closing valves, the operator can withdraw water from different depths to compensate for changing water elevation, to avoid ice cover, or to select the depth from which the best-quality water can be withdrawn.

Pump Intakes

Pump intakes are used where it is not possible to construct a facility that will allow water to flow from the source by gravity. One design, illustrated in Figure 3-8, has an intake pump and suction piping mounted on a dolly that runs on tracks. As the water level rises or falls, sections of discharge piping are removed or added and the rail car is moved accordingly.

Infiltration Galleries

An infiltration gallery is an intake that collects seepage water. It is usually located next to a lake or river where it can pick up seepage from the surrounding sand. A "buried intake" (instead of an open intake) can similarly be located in the bottom of a lake or stream, but it can be used only if the bottom sand is porous enough and has the proper gradation to provide a

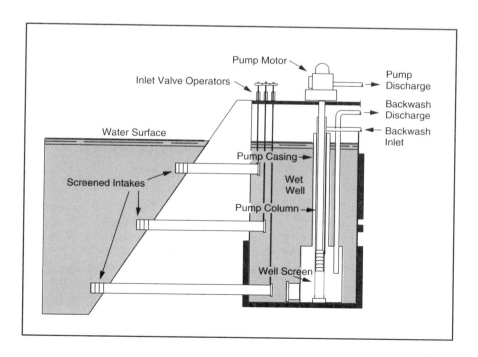

FIGURE 3-7
Intake structure with multiple inlet ports

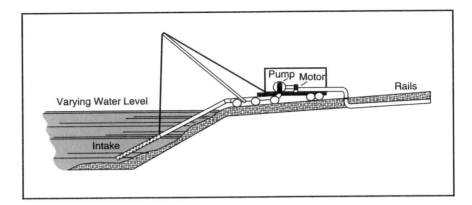

FIGURE 3-8
Movable pump intake

reasonable flow rate. There must also be enough wave action or flow across the bottom to prevent the accumulation of silt, which would plug the sand surface.

Where the proper conditions are present for installing an infiltration system, infiltration galleries will usually provide much better water quality than water taken directly from the source because the natural straining action of the sand eliminates debris and reduces turbidity levels. They also have the advantages of not freezing or icing, and they are not as vulnerable to damage.

Surface Supply Operating Problems

Surface water sources can have a number of physical, chemical, and biological problems that are troublesome for the water system operator.

Stream Contamination

Streams in particular are at the mercy of all upstream users of the water and land within the drainage basin. Rivers of commerce such as the Ohio and Mississippi are often in the news because of spills from barges, leaks from tank facilities, broken pipelines, accidental industrial spills, and other assaults on stream water quality. Utilities using particularly vulnerable streams as a water source generally depend on warning systems to give advance notice of problems approaching their intake. Some streams also have occasional periods when water quality is especially poor as a result of natural causes, such as heavy spring runoff from a swampy area.

Any water system that is vulnerable to poor-quality episodes should construct adequate storage facilities. These facilities are then used when the intake is shut down due to poor-quality water passing the intake point.

Lake Contamination

Lakes and reservoirs are also vulnerable to natural and human contamination. One problem encountered by some water systems using reservoirs in agricultural areas is the occasional occurrence of nitrate levels exceeding state and federal standards. The nitrates are caused by runoff from farmlands located in the drainage basin; they may persist for several months each year.

Excessive growth of algae and aquatic weeds in reservoirs is also quite common, particularly in warmer climates. This is usually caused by high levels of nutrients in the water, but the problem can be minimized by proper treatment.

A recent threat introduced to the Great Lakes by commercial shipping is the zebra mussel. This organism grows profusely in intake structures and can quickly reduce water flow if not controlled. Another mussel, the quagga, has been found to cause similar problems. These organisms are gradually being transported to other surface water impoundments by boats and birds. It appears that they will probably become established as permanent nuisance organisms in most waters of North America.

Control of aquatic plants, prevention of lake stratification, control of mussels, and other treatment of source water are detailed in *Water Treatment*, also part of this series.

Icing

Operators in cold-weather areas often face the additional operational problems of frazil ice or anchor ice. When the water is almost at the freezing point and is rapidly being cooled, small, disk-shaped frazil ice crystals will form and be distributed throughout the water mass. When the frazil crystals are carried to the depth of a water intake, they may adhere to the intake screen and quickly build up to a solid plug across the opening.

Anchor ice is slightly different in that it is composed of sheetlike crystals that adhere to and grow on submerged objects. Although experience varies at different locations, most systems experience ice blocking only at night. The conditions that have generally been found to be favorable to ice crystal formation are

- clear skies at night (because there will be high heat loss by radiation)

- air temperature of less than 19.4°F (–7°C)
- daytime water temperature not greater than 32.4°F (0.2°C)
- winds greater than 10 mph (16 km/h) at the water surface

Icing of an intake will usually be indicated by a falling level in the intake well. It is generally best to stop using the intake immediately when icing is noticed because the blockage will only get worse. If sufficient storage is available or another intake can be used, the easiest method of ridding the intake of ice is to just wait and let it float off within a few hours.

Water systems that frequently experience icing have tried such things as providing the intake with piping to backflush the line with settled water or blowing the line with compressed air or steam. If more than one intake is available, use of individual intakes can be periodically switched to allow any accumulated ice to melt and float away from the intake opening.

Methods of constructing new intake structures to minimize ice formation include using very widely spaced screening or no screen at all and using nonferrous materials such as fiberglass for the intake pipe and structures. Many intakes are also designed to keep the entrance velocity very low by using large bell-shaped structures or several inlets.

Stream source intakes are not ordinarily deep enough to allow water to be taken from a depth beneath the supercooled level, so multiple intake points and use of heat may be necessary to prevent icing.

Evaporation and Seepage

Storage impoundments that have large exposed surface areas commonly lose 6 to 8 ft (around 2 m) of water a year through evaporation. In addition, there are seepage losses through the bottom and sides of the impoundment. These losses cannot economically be controlled in large reservoirs, so the reservoirs are generally designed larger than required to compensate for the expected loss.

In small excavated reservoirs, seepage can be controlled by lining the impoundment. Evaporation can also be controlled by covering a reservoir with floating plastic sheets (Figure 3-9) or applying a thin layer of a liquid chemical to the surface.

Siltation

All streams carry sediment. Most of this sediment is deposited when water loses velocity upstream from a dam. The rate of siltation is generally a function of both the type of soil in the area and how well the land is protected

FIGURE 3-9
Reservoir cover
installed on
Garvey Reservoir
at Monterey Park,
Metropolitan
Water District of
Los Angeles, Calif.

Courtesy of Burke Environmental Products, a Division of Burke Industries, Inc.

by vegetation. In some locations, siltation is so rapid that most of the capacity of a dam reservoir is lost within a few years after it is constructed.

The problem of siltation can never be completely solved, but control of the watershed can minimize it. Enforcement of rules governing good practice in farming, logging, road construction, and other operations that disturb the land will help reduce the amount of sediment entering tributary streams.

Creation of artificial wetlands for small streams can also help reduce siltation. The reduction in the stream's velocity caused by the wetlands allows the bulk of the sediment load to drop out in the wetlands before the water reaches the impoundment.

When the capacity of a reservoir has been decreased by siltation, there are three general options: the silt can be removed by dredging; the reservoir can be drained and the silt excavated; or a new reservoir can be constructed. Draining and excavating are only an option when the water system has other water sources available, and constructing a new reservoir is only possible if the necessary additional land is available. A cost analysis of the available options must be made to decide how best to restore the lost reservoir capacity.

Selected Supplementary Readings

Manual of Individual Water Supply Systems. 1982. US Environmental Protection Agency. Office of Drinking Water. EPA-570/9-82-004. Washington, D.C.: US Government Printing Office.

Manual of Instruction for Water Treatment Plant Operators. 1975. Albany, N.Y.: New York State Department of Health.

Manual of Water Utility Operations. 8th ed. 1988. Austin, Texas: Texas Water Utilities Association.

Water Quality and Treatment. 4th ed. 1990. New York: McGraw-Hill and American Water Works Association (available from AWWA).

CHAPTER 4

Emergency and Alternative Water Sources

It is important that service to a public water system be as close to continuous and uninterrupted as possible. Although it is an inconvenience to the public to be without water for any length of time, the most important reason for maintaining continuity of service is for protection of public health. Any loss of system pressure or interruption in service is an opportunity for microbiological or chemical contamination to enter the water system. In addition, lack of water for sanitary purposes for more than a short period of time can create problems with disease. The possibility of water contamination or loss of the normal water source requires that every water utility develop an emergency response plan. Figure 4-1 illustrates a disaster effects matrix for a typical utility.

An emergency response plan should be developed by each utility in cooperation with neighboring utilities and with the appropriate local and state agencies. These other agencies include local police, fire, and emergency or disaster preparedness departments, as well as the same groups at the county and state level. The purpose of this chapter is not to address the whole emergency planning program but only to study possible emergencies involving source water. Alternative water sources for nonpotable use are also discussed.

Causes of Source Disruption

In general, calamities that can disrupt the operation of a water source fall into the following two categories: natural disasters, including earthquakes

**FIGURE 4-1
Disaster effects
matrix**

Figure continued next page

System Components — Likely damage, loss, or shortage due to hazards	Earth-quakes	Hurri-canes	Tornados	Floods	Forest or brush fires	Volcanic eruptions	Other severe weather	Water-borne disease	Hazard-ous material	Struc-ture fire	Con-struction accidents	Transpor-tation accidents	Nuclear	Vandals, riots, strikes
Administration/operations														
Personnel	■■■	■■■	■■	■■	■■	■	■	■			■	■	■■	■■■
Facilities/equipment										■■				■■
Records														■
Source water														
Watersheds/surface sources		■■		■■■■	■	■		■■■■	■■■■				■■■	■■
Reservoirs and dams	■	■■												
Groundwater sources														■
Wells and galleries	■	■	■	■		■	■							
Transmission														
Intake structures	■■■			■■■■	■	■	■				■■■	■■■		■
Aqueducts	■■	■■■■	■■	■■	■■■■	■	■			■				■■
Pump stations	■■	■	■											
Pipelines, valves	■													
Treatment														
Facility structures	■■■	■■■	■■■	■■■	■■■■	■	■■			■■■■	■■■	■■■	■	■■■■
Controls	■■	■	■							■	■	■	■	■
Equipment	■													
Chemicals									■					
Storage														
Tanks	■■■	■■■	■		■	■	■■■	■	■	■		■■	■	■■
Valves														
Piping														
Distribution														
Pipelines, valves	■■■	■■■	■	■■	■■	■	■■	■	■	■■	■■	■		■■■
Pump or PRV stations														
Materials														

System Components — Likely damage, loss or shortage due to hazards	Earth-quakes	Hurri-canes	Tornados	Floods	Forest or brush fires	Volcanic eruptions	Other severe weather	Water-borne disease	Hazard-ous material	Struc-ture fire	Con-struction accidents	Transpor-tation accidents	Nuclear	Vandals, riots, strikes
Electric power														
Substations	■	■	■	■	■		■			■	■	■	■	■
Transmission lines	■	■	■	■	■		■			■	■	■		■
Transformers	■	■	■	■	■	■	■			■	■	■		■
Standby generators				■						■				■
Transportation														
Vehicles	■	■	■	■	■		■			■	■	■		■
Maintenance facilities		■	■	■	■		■			■		■		■
Supplies	■	■		■	■	■	■				■	■		■
Roadway infrastructure														
Communications														
Telephone	■	■	■	■	■		■			■	■	■		■
Two-way radio	■		■	■	■	■								■
Telemetry														■

FIGURE 4-1
Disaster effects matrix (continued)

(Figures 4-2 and 4-3), floods, hurricanes, tornados, forest fires, landslides, snow and ice storms, and tidal waves; and human activities, including vandalism, explosions, strikes, riots, terrorism, warfare, and water contamination caused by leaks, spills, or dumping of hazardous materials. In addition, there is always the possibility of major equipment failure and breakdown caused by human error.

For whatever reason the water source is disrupted, the water can effectively become unusable because it could be so contaminated that it is far beyond the ability of available equipment to treat it to acceptable quality, or because the water might no longer be available from the source as a result of complete mechanical failure of the treatment and distribution system.

Source Contamination

Contamination of source water can be caused by microbiological, chemical, or radiological agents. Both groundwater and surface water sources are vulnerable. Some relatively common examples of contamination incidents are

- a pipeline that has broken upstream of a surface water intake, allowing several hundred gallons (liters) of fuel oil to be spilled into the river

FIGURE 4-2 A 400,000-gal (1.5 million-L) steel tank uplifted during the 1992 Landers, Calif., earthquake

Source: M.J. O'Rourke, RPI, NCEER.

FIGURE 4-3
Flocculator/clarifier
center mechanism
damaged during
1989 Loma Prieta
earthquake

Source: D.B. Ballantyne.

- a tank truck that has overturned on a bridge and has spilled a quantity of toxic chemical into the river
- a spill of chemical solvent on the ground near a well that has penetrated to the aquifer and is now showing up in the well at increasing concentrations

When contamination of a water source occurs, the first thing that must be determined is the length of time for the contamination episode. In the above examples, the tank truck spill will probably pass the river intake in a matter of hours. The fuel oil spill could last for several days. The aquifer contamination, on the other hand, could be permanent; if it will pass by naturally or can be cleaned up, the process may take years.

When a source water becomes contaminated a utility can stop drawing water and operate on stored water until the episode is over, change to an alternative water supply until the episode is over, or treat the contaminated water.

Most chemical contaminants in water can be reduced to acceptable levels with the proper type of treatment, such as aeration or carbon adsorption. Microbiological contaminants can generally be controlled by proper coagulation, sedimentation, and filtration treatment plus heavy doses of disinfectant. However, the equipment required for this special treatment may not be required for normal source treatment. Water system operators should therefore consider the contamination problems that might occur and have the special equipment available or make plans to obtain it quickly if needed. Additional details on water treatment methods for contaminant removal are included in *Water Treatment*, also part of this series.

Loss of Water Source

A complete (temporary or permanent) loss of a water source could occur when an earthquake has damaged all of a water system's wells or when a flash flood on a river has destroyed a system's intake facilities.

If a water system's usual water source is no longer available because of mechanical disruption, the first decision is whether the facility is repairable or if temporary facilities can be installed. For example, the damaged wells may not be repairable, in which case new wells may have to be drilled, and that could take months. If a surface water intake is destroyed, it may be possible to install a temporary pipeline or bring in portable pumps. These tasks might be accomplished in a few days.

Short-Term Options

The following short-term options should be considered when a water source is damaged or destroyed:

- If a limited supply of water is still available, commence conservation or rationing.
- Supply customers with water from tank trucks.
- Supply customers with bottled water.
- Draw water from adjoining water systems.

Long-Term Options

The following long-term options should be considered when a water source is damaged or destroyed:

- Drill new wells.
- Construct a new surface water source.
- Clean up the source of contamination.
- Install a connection to purchase water from another water system.
- Impose permanent conservation requirements.
- Reuse wastewater for nonpotable requirements.

- Install a dual potable–nonpotable water system.
- Construct new raw-water storage, such as a new reservoir.
- Perform aquifer recharge.

Evaluating the Options

There are many factors that must be considered and information that must be gathered before a decision can be made on which emergency options to choose. Some of these concerns are as follows:

Technical and logistical feasibility

- What procedures are required to implement the option?
- Is the required technology available and properly developed?
- How much water can the option provide?
- Will the option meet only the system's priority water needs?
- Can it meet the system's current total water needs?
- Can it be expanded to meet future community water needs?
- How quickly can the option be made operational?
- What equipment and supplies are needed?

Reliability

- How reliable is the option?
- Does it require special operation and maintenance skills?

Political considerations

- What administrative procedures are required to implement the option?
- Is property ownership a problem?
- Will the option be acceptable to the public?

Cost considerations

- What is the initial investment required?
- What will the operating costs be?
- Who will bear the cost of the design, construction, and operation of the option?

Alternative Water Sources

Review of the many alternatives and considerations that must be made after an emergency emphasizes why *advanced* planning should be done to determine what options a water system has in the event of a disaster. Most systems do not have readily available alternative sources, so some thought and planning must be given to how water can be provided if the major water source is contaminated or must be out of service for a prolonged period of time. If at all possible, an alternative water source should be provided in advance, rather than after the emergency happens. Some typical alternative sources that should be considered are discussed below.

Surface water systems

- Provide two or more intakes at different locations in the lake or stream.
- Provide intakes in more than one water source, if possible.
- Construct wells for backup of the surface source.

Groundwater systems

- Provide enough wells to meet demand when one or more wells are out of service.
- Locate wells in different aquifers or far enough apart in the same aquifer so that contamination affecting one well will not affect other wells.
- Provide a surface water emergency source.

All systems

- Establish interconnection with other water systems, with sufficient capacity to meet at least minimum needs.

Many potential emergencies can be averted or minimized by making preparations in advance. For instance, good security at water facilities can reduce vandalism; providing standby power eliminates complete dependence on commercial electric power; and locating all facilities far enough away from flood plains can prevent many contamination problems. Besides averting problems that can possibly be prevented, water system operators should be prepared to act swiftly and efficiently in the event of an emergency.

Intersystem Connection

In urban and suburban areas, it is quite common to find water utilities that have individual water sources but distribution systems that are located close to one another at some points. Even without prior planning, emergency connections can be made between the distribution systems using fire hoses or temporary piping. Unfortunately, this is time consuming in an emergency and will usually provide only a fraction of the water needed.

Efforts should be made in advance to interconnect adjoining distribution systems or, if practical, the raw-water transmission systems themselves. If water mains are large enough at adjacent points and water pressure is about the same, the interconnection can be relatively simple.

The managers of adjoining water systems should carefully plan for either system to furnish water to the other in an emergency. A satisfactory connection will require some piping to connect with feeder mains of sufficient size to provide the needed amount of water in either direction. It may also be necessary to install pressure-reducing valves and a booster station if system pressures are not compatible. It may also be desirable to provide meters for flow in either direction, for purposes of either reimbursement or just water accountability.

Emergency Provision of Water

Tank trucks and bottled water can provide emergency sources of water. Rationing can be used if the supply is reduced but the distribution system is still operating.

Tank Trucks

If a water distribution system suffers massive failure and no alternative source is available, nonpiped water must be used. Tank trucks (Figure 4-4) can be used for this purpose, and every utility should maintain a record of where suitable units are available. The utility must be assured of the chemical and biological suitability of the trucks and must take precautions such as ensuring that adequate disinfection is maintained.

If the water utility does not supply the driver, it must have assurance that the driver uses recognized basic sanitary precautions and, most importantly, that he or she draws water from an acceptable water source. If water is to be taken from a neighboring utility, communication and monitoring are necessary to ensure that the water-filling point is satisfactory to both utilities.

Bottled Water

Bottled water is also frequently made available to customers when a water system fails or the water is contaminated. The demand will quickly exceed the normal supply, so arrangements must be made for supplemental

FIGURE 4-4
Emergency
potable water
provided by
tanker truck in
Kauai, Hawaii

Source: Ray Sato

supplies. In some cases, local dairies and beverage bottlers will provide water in milk cartons or bottles during an emergency.

In general, the public's use of bottled water for drinking and some cooking has been steadily increasing. Public perception that tap water may not be safe has been part of the reason for the increased use of bottled water, as well as taste, odor, and other quality objections. It is the responsibility of water utility operators and managers to provide water that is safe and palatable. Safety can never be compromised, and palatability should be a goal held constantly in mind, if not already achieved.

Bottled water can definitely be helpful in certain emergency or disaster situations. This does not have to imply the need to buy expensive brands or import water from great distances. Beverage companies in nearby communities not affected by the emergency can be called on to help. They will usually have a supply or ready source of containers and also the distribution facilities in the form of special trucks.

It is important to note that bottled water cannot be used as a substitute for making a public water supply meet the standards of regulatory agencies.

Water Rationing

For some disasters, a reduced supply of water results, but the distribution system is still functioning. Typical situations include the following:

- The system can reestablish the water source soon and has enough water in storage to supply basic customer needs if use is drastically reduced.
- An interconnection has been established with another utility, but it provides only enough water to supply basic needs.
- The water system must operate at very low pressure because of pump failure or other problems.

In these cases, a severely reduced demand can help alleviate the crisis. The utility's conservation plan or emergency response plan should have all categories of water use rated. This situation is comparable to that faced by communities with severe or critical drought conditions. Table 4-1 is an example of a plan prepared for use in such situations.

Water Reuse

As populations grow and potable water becomes scarcer in some areas, increasing consideration will have to be given to water reuse. The three ways in which water reuse is currently considered are indirect reuse, direct reuse, and by providing dual-water systems.

Indirect Reuse

Indirect potable reuse — the use of water from streams that have upstream discharges of wastewater — has been in existence for a very long time. For instance, the discharges from individual septic systems infiltrate into aquifers, and at some later time the same water is withdrawn from wells. Sewage and industrial wastes discharged to rivers are diluted and somewhat treated by natural actions, and the water is later withdrawn by downstream communities for water supply use. Every day millions of people worldwide consume water that is partially reuse water.

At present there is no alternative for this situation, and it is likely to increase as urban populations increase. As long as both the dischargers and receivers are aware of what is happening and treat the wastewater and the potable water to the best of their ability, the situation is definitely better than unplanned reuse.

Recharging of groundwater by surface spreading or direct injection of reuse water is being practiced on a large scale in some areas. The

TABLE 4-1 Example of a drought or emergency conservation plan

Stage	Degree of Drought or Emergency	Consumption Reduction Goal, %	Public Information Action	Public-Sector Action	User Restrictions	Penalties for Noncompliance
I	Minor	10	Explain drought or emergency conditions. Disseminate technical information. Explain other stages and possible actions. Distribute retrofit kits at central depots. Request voluntary reduction.	Increase enforcement of hydrant-opening regulations. Increase meter-reading efficiency and meter maintenance. Implement intensive leak detection and repair program.	Voluntary installation of retrofit kits. Restriction of outside water use for landscape irrigation, washing cars, and other uses.	Warning
II	Moderate	15–18	Use media intensively to explain emergency. Explain restrictions and penalties. Explain actions in potential later stages. Request voluntary reduction.	Reduce water usage for main flushing, street flushing, public fountains, and park irrigation.	Mandatory restriction on all outside uses by residential users, except landscape irrigation. Prohibit unnecessary outside uses by any commercial users.	1. Warning 2. House call 3. Installation of flow restrictor 4. Shutoff and reconnection fee

Table continued next page

TABLE 4-1 Example of a drought or emergency conservation plan (continued)

Stage	Degree of Drought or Emergency	Consumption Reduction Goal, %	Public Information Action	Public-Sector Action	User Restrictions	Penalties for Noncompliance
III	Severe	25–30	Public officials appeal for water use reduction. Explain actions and consequences of emergency.	Prohibit all public water uses not required for health or safety.	Severely restrict all outside water use. Prohibit serving water in restaurants. Prohibit use of water-cooled air conditioners without recirculation.	Same as Stage II
IV	Critical	50 or more	Same as Stage III	Reduce system pressure to minimum permissible levels. Close public water-using activities not required for health or safety.	Prohibit all outside water use and selected commercial and industrial uses. Terminate service to selected portions of system as last extreme measure.	Same as Stage II

groundwater can then be pumped for nonpotable uses, in which case demand is reduced on the potable water aquifer.

Direct Reuse

A large potential source of additional water for water-short areas of the world is the direct reuse of wastewater for purposes not requiring potable water. Wastewater may include treated municipal sewage, industrial cooling and process water, and agricultural irrigation runoff water. Any water that is used again after being used by any consumer is classified as reclaimed water. At the present time, the principal use of reclaimed water is for agricultural irrigation. Figure 4-5 shows a typical irrigation concept using reclaimed water. For decades, certain areas of the United States, such as the southwestern states, have successfully been applying reclaimed water to golf courses, orchards, and nonedible crops, and using it for flushing toilets (Figure 4-6).

The use of wastewater for potable purposes is not generally considered acceptable at the present time. The American Water Works Association (AWWA) opposes the direct use of reclaimed wastewater as a potable water supply source until more research can be completed. It is the general feeling that the treatment technology and monitoring techniques are not yet available to ensure that all processed wastewater is completely safe for human consumption.

Dual-Water Systems

One method of achieving water reuse is to deliver potable water and nonpotable water to consumers through separate distribution systems. This

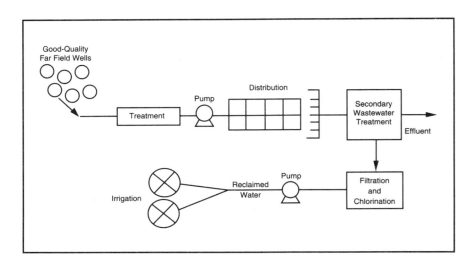

**FIGURE 4-5
Irrigation concept
using reclaimed
water**

FIGURE 4-6 El Tovar Lodge in Grand Canyon Village, Ariz., the site of the first US dual-distribution system in the United States (reclaimed water is used for landscaping and toilet flushing)

Source: Daniel A. Okun

is generally referred to as using a dual-water system. Use of a dual system is also a possible solution where potable water is in very limited supply but lower-quality, nonpotable water is more plentiful. The concept is not new and has been used occasionally by water systems having unusual source water conditions.

Analyzing water demand, especially domestic use, gives a better picture of the potential for the great savings of potable water through the use of dual-water systems. Within the home, high-quality potable water is needed for drinking, cooking, bathing, laundry, and dishwashing. The other uses, primarily toilet flushing, do not necessarily require high-quality water.

Exterior water use as a percentage of total use varies dramatically with the seasons. On an annual basis, exterior use averages 7 percent in Pennsylvania and 44 percent in California. Because high-quality water is not required for exterior residential uses, the demand can be satisfied by the use of nonpotable water.

There are a number of areas in the United States where dual-distribution systems are in use; for example, Tucson and Phoenix, Ariz.; Irvine, Calif.; and Colorado Springs and Denver, Colo.

FIGURE 4-7 Reclamation plant in St. Peterburg, Fla., that incorporates secondary treatment, filtration, and chlorination

Source: St. Petersburg Department of Public Utilities.

Probably the most extensive system is in St. Petersburg, Fla.; Figure 4-7 shows its water reclamation plant. The local water supply there was exhausted in 1928 because of saltwater intrusion into the water system wells. Since that time, other well fields have been developed, some as far as 50 mi (80 km) away, but the city is still expected to have a severe shortage of potable water after the turn of the century. A great portion of the water use in the city is for irrigation of domestic lawns, as well as municipal and institutional lawns and green spaces. A very detailed financial study was undertaken comparing the cost of various alternatives. Overall savings after installing a dual-water system — which include the savings from not having to develop more potable-water source capacity — are more than 6 percent. An added benefit arising from the nature of the dual system is the savings on fertilizer costs. For example, the amount of nitrogen, phosphorus, and potassium in the reuse water reduced fertilizer costs on one golf course alone by $6,000 annually.

It is likely that more dual systems will be constructed in the future, but public health consideration must be paramount. Improper use of nonpotable water must be controlled, cross-connections strictly monitored, and responsibility for the nonpotable system clearly established.

Selected Supplementary Readings

Guide to Ground-Water Supply Contingency Planning for Local and State Governments. 1991. Washington, D.C.: US Environmental Protection Agency, Office of Drinking Water.

Manual M19, Emergency Planning for Water Utility Management. 1994. Denver, Colo.: American Water Works Association.

Manual M24, Dual Water Systems. 1994. Denver, Colo.: American Water Works Association.

Minimizing Earthquake Damage: A Guide for Water Utilities. 1994. Denver, Colo.: American Water Works Association.

CHAPTER 5

Use and Conservation of Water

Water is our most important natural resource, but up until the last two decades, it was generally taken for granted by most customers. Now, as a result of growing environmental awareness and increases in the cost of water service, consumers are beginning to look at water, and at water utilities in general, from a different perspective.

Why has the cost of water increased? One of the primary reasons is the added cost of complying with the Safe Drinking Water Act (SDWA) and other environmental regulations. Drinking water is being treated and tested as never before. Increased health-related concerns have caused water utilities to improve the quality of their monitoring and treatment, resulting in increased capital and operating costs.

Any action of a water utility that impacts the environment in any way must undergo public scrutiny. For example, unless the utility can fully justify constructing a new dam, enlarging a reservoir, installing a pipeline, or putting up a standpipe or pumping station, it is likely to be denied permission to do so. All of these demands put pressure on the utility and on water system operators to make sure that every gallon of water delivered to customers costs no more than it has to, meets state and federal health requirements, is palatable, and is used wisely by the public, business, and industry.

Water Use

Total water use has declined in the United States since 1980 because of reductions in use by power producers, industry, and farmers. The amount used for public water supply, however, has continued to increase.

Major Uses of Water

It is estimated that approximately 408 bil gal (1,544 GL) of fresh water are withdrawn each day in the United States for all uses.

In 1990 the five principal uses for this water and their daily amounts were as follows:

Public water supplies	39 bil gal (148 GL)
Thermoelectric power	195 bil gal (738 GL)
Irrigation	137 bil gal (518 GL)
Other industrial direct withdrawal	30 bil gal (113 GL)
Rural water use	8 bil gal (30 GL)

Figure 5-1 shows the progression of these water uses for 1950 through 1990.

Public Water Supply Use Categories

The total use by public water supplies averages out nationally to about 183 gallons per capita per day (gpcd) (693 L/d per capita). All public water supply uses can be grouped into four categories, as shown in Figure 5-2, which also shows the percentages of use for each category.

If these percentages are multiplied by the 1990 average per capita use, the average use by category is as follows:

Domestic use	105 gpcd (397 L/d per capita)
Commercial use	28 gpcd (106 L/d per capita)

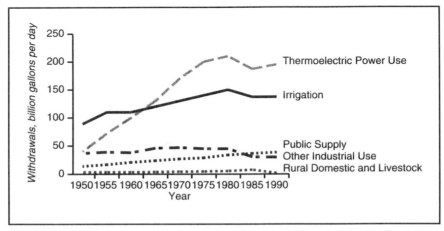

**FIGURE 5-1
Trends of
estimated water
use in the United
States, 1950–1990**

Adapted from Solley, Pierce, and Perlman (1993).

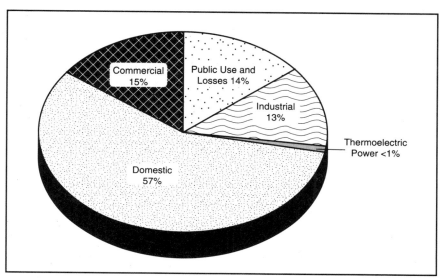

FIGURE 5-2   US national average public water supply use

Source: Solley, Pierce, and Perlman (1993).

| Industrial use from public water systems | 25 gpcd (95 L/d per capita) |
| Public use | 26 gpcd (98 L/d per capita) |

The actual use rate in any community may, of course, vary widely from this average. Major factors that affect local use are water availability, climate, and the number of water-using industries.

Domestic Use

Activities using water in and around the home are termed domestic water uses. The amount of water used varies with factors such as the age and number of occupants, income level, and geographic location. Table 5-1 shows an analysis of water use by a typical family of four persons living in a single-family house. Table 5-2 shows flow rates for various water-using fixtures. These tables indicate that a relatively small percentage of the water used by residential customers is used for drinking, cooking, and kitchen use. The largest uses are for toilet flushing, bathing, laundering, and lawn watering.

TABLE 5-1 Typical urban water use by a family of four

Type of Household Use	Daily Use				
	Per Family			Per Capita Use, gpcd (L/d per capita)	
	Amount of Water Used, gpd (L/d)		%		
Drinking and water used in kitchen	8	(30)	2	2.00	(7.6)
Dishwasher (3 loads per day)	15	(57)	4	3.75	(14.2)
Toilet (16 flushes per day)	96	(363)	28	24.00	(90.8)
Bathing (4 baths or showers per day)	80	(300)	23	20.00	(75.7)
Laundering (6 loads per week)	34	(130)	10	8.50	(32.2)
Automobile washing (2 car washes per month)	10	(38)	3	2.50	(9.5)
Lawn watering and swimming pools (180 hours per year)	100	(380)	29	25.00	(94.6)
Garbage disposal unit (1 percent of all other uses)	3	(11)	1	0.75	(2.8)
Total	346	(1,310)	100	86.50	(327.4)

Reprinted with permission from The Water Encyclopedia. 2nd ed., 1990.
Copyright CRC Press, Inc., Boca Raton, Fla.

Commercial Use

Most public water systems also serve a variety of commercial water users such as

- motels
- office buildings
- shopping centers
- laundries
- car washes
- service stations

TABLE 5-2   Flow rates for certain plumbing, household, and farm fixtures

Location	Flow Pressure,*		Flow Rate,	
	psi	(kPa)	gpm	(L/min)
Ordinary basin faucet	8	(55)	2.0	(7.6)
Self-closing basin faucet	8	(55)	2.5	(9.5)
Sink faucet, ⅜ in. (9.5 mm)	8	(55)	4.5	(17.0)
Sink faucet, ½ in. (12.7 mm)	8	(55)	4.5	(17.0)
Bathtub faucet	8	(55)	6.0	(22.7)
Laundry tub faucet, ½ in. (12.7 mm)	8	(55)	5.0	(18.9)
Shower	8	(55)	5.0	(18.9)
Ball-cock for closet	8	(55)	3.0	(11.3)
Flush valve for closet†	15	(103)	15–40	(57–151)
Flushometer valve for urinal	15	(103)	15.0	(56.8)
Garden hose (50 ft [15 m], ¾-in. [19-mm] sill cock)	30	(207)	5.0	(18.9)
Garden hose (50 ft [15 m], ⅝-in. [15.9-mm] outlet)	15	(103)	3.33	(12.6)
Drinking fountains	15	(103)	0.75	(2.8)
Fire hose 1½-in. (38-mm) hose and ½-in. (12.7-mm) nozzle	30	(207)	40.0	(151.4)

*Flow pressure is the pressure in the supply near the faucet or water outlet while the faucet or water outlet is wide open and flowing.
†There is a wide range of flow rates in toilet flush valves due to variations in design.

Reprinted with permission from The Water Encyclopedia. *2nd ed., 1990. Copyright CRC Press, Inc., Boca Raton, Fla.*

- airports
- swimming pools

In addition to water used for drinking and sanitary purposes by employees and customers, many commercial users also have special water needs, such as for cooling, humidifying, washing, ice making, vegetable and produce watering, ornamental fountains, and landscape irrigation.

Industrial Use

Water use by industries varies widely. It depends on the type of industry, water cost, wastewater disposal practices, types of processes and equipment, and local water conservation and reuse practices. In general, industries that use large quantities of water are located in communities where water quality is good and the cost is reasonable.

Historically, major industrial water users (such as the steel, petroleum products, pulp and paper, coke, and power industries) have provided their own water supplies. Smaller industries and industries with low water usage generally purchase water from public systems.

Water used by industry is usually measured in terms of the amount of product produced. Table 5-3 lists water requirements for some of the industries that use large quantities of water.

Public Use

Municipalities and other public entities provide services that require varying amounts of water. Some typical public uses are

- public parks, golf courses, swimming pools, and other recreational areas
- municipal buildings
- fire fighting
- public works uses such as street cleaning, sewer flushing, and water system flushing

Although the total annual volume of water required for fire fighting is small, the rate of flow required during a fire can be very large. For a large fire, it can be the largest single demand a water system will ever need to meet. For example, a community that uses an average of 1 mgd (4 ML/d) might reasonably expect a fire flow demand of 2,500 gpm (9,500 L/min) for up to 10 hours. The total amount of water used in the fire would be 1.5 mil gal (5.7 ML) in addition to normal water system demands. This is over twice the average daily flow.

Variations in Water Use

There are a number of factors that can cause either more or less water to be used. Some of the major ones are

- the time of day and the day of the week
- the climate and the season of the year
- the type of community (residential or industrial) and the economy of the area
- system water pressure
- the presence or absence of meters
- the quality of the water

TABLE 5-3 **Water requirements in selected industries**

Industry	Water Required per Unit of Product Noted	
	gallons	litres
Food and Beverage		
Meatpacking, per ton (0.9 metric ton) carcass weight	7,200	27,200
Dairy products, per ton (0.9 metric ton) milk processed	1,700	6,400
Canned fruits and vegetables, per ton (0.9 metric ton) vegetables canned	19,700	74,600
Frozen fruits and vegetables, per ton (0.9 metric ton) vegetables frozen	22,500	85,200
Malt beverages, per 1,000 gal (3,785 L) beer and malt liquor	50,000	190,000
Cane sugar, per ton (0.9 metric ton) cane sugar	28,100	106,400
Beet sugar, per ton (0.9 metric ton) beet sugar	33,100	125,300
Petrochemical		
Plastic materials and resins, per ton (0.9 metric ton) plastics	47,000	177,900
Paints and pigments, per 1,000 gal (3,785 L) paint	13,200	50,000
Nitrogenous fertilizers, per ton (0.9 metric ton) fertilizer	28,500	107,900
Phosphatic fertilizers, per ton (0.9 metric ton) fertilizer	35,600	134,700
Petroleum refining, per 1,000 gal (3,785 L) crude petroleum input	44,000	166,500
Synthetic rubber, per ton (0.9 metric ton) synthetic rubber	110,600	418,600
Textile Products		
Textile mills, per ton (0.9 metric ton) textile fiber input	69,800	264,200
Wood Products, Pulp, and Paper		
Pulp and paper mills, per ton (0.9 metric ton) paper	130,000	492,000
Paper converting, per ton (0.9 metric ton) paper converted	6,600	25,000

Adapted from Kollar and MacAuley (1980).

Table continued next page

TABLE 5-3   Water requirements in selected industries (continued)

Industry	Water Required per Unit of Product Noted	
	gallons	*litres*
Metals and Metal Production		
Steel, per ton (0.9 metric ton) steel net tons (0.9 metric tons)	62,600	236,900
Iron and steel foundries, per ton (0.9 metric ton) ferrous castings	12,400	46,900
Primary copper, per ton (0.9 metric ton) copper	106,000	401,200
Primary aluminum, per ton (0.9 metric ton) aluminum	98,300	372,100
Manufacturing		
Automobiles	36,500	138,200

Adapted from Kollar and MacAuley (1980).

- connection to a sewer system or to individual septic systems
- the condition of the water system

Time and Day

The time of day and the day of the week can have a profound effect on water use. In most communities, water use is lowest in the early morning hours when most people are asleep. On a typical day, water use rises rapidly when customers awake and then typically levels out through the day. Usage then increases in late afternoon, peaks during the evening, and drops rather quickly around 10:00 p.m. The changing hourly rate of water use for a typical day is shown in Figure 5-3.

Different days of the week usually show different total water use, with day-to-day patterns depending on the habits of the community. Some water systems can see a significant increase in water use on Monday because it is historically "wash day" for many households. Water systems supplying industries that do not operate on weekends will often have much lighter water use on Saturday and Sunday.

Climate and Season

Water use is usually highest during summer months, particularly in warm, dry climates. More water is used for bathing, lawn and garden sprinkling, and other outside activities.

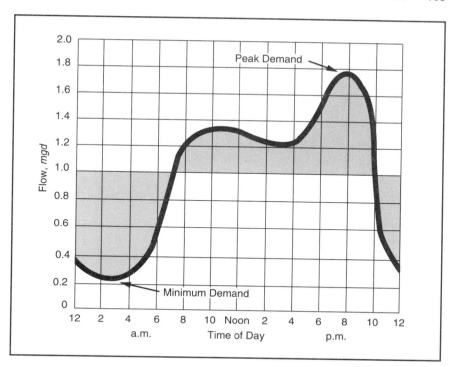

FIGURE 5-3
Typical daily flow chart showing peak and minimum demands

Most systems have relatively low water use in winter when there is no garden sprinkling or other outside activities. However, in extremely cold weather, consumers might run water faucets continuously to prevent water pipes from freezing.

Home air coolers that operate by evaporating water, also referred to as swamp coolers or evaporative coolers, are used extensively in homes in hot, dry climates. They are also widely used in large commercial and industrial facilities. The demand for water to support air coolers generally coincides with periods of highest demand for lawn and garden sprinkling.

Type of Community

Residential communities will use less water per person than highly commercialized and industrialized communities. The type of housing that is most common in the service area will also affect use. For example, areas with individual homes on large lots that have gardens and lawns will have a much higher water use per person than areas with multiple-family dwellings such as townhomes, condominiums, and apartment complexes. Household appliances, including automatic washing machines, dishwashers, and garbage disposals, all tend to use more water and will increase overall water use where they are installed. Economically depressed areas will generally have a lower per capita water use.

Water Pressure

The amount of water used by consumers also increases with increased system pressure. A gauge pressure of 25 to 50 psig (170 to 345 kPa) is considered normal. By increasing a 25-psig (170-kPa) service pressure to 45 psig (310 kPa), a utility can expect to increase water use by as much as 30 percent.

Metering

Most water utilities have meters installed on all water services, but a few systems continue to charge residential customers a flat rate. It has been found that flat-rate customers may use as much as 25 percent more water than is used by customers who have to pay for the exact amount of water they use. This excess water usage can be caused by unnecessary and unrestricted sprinkling, and by obvious wastage, i.e., letting a faucet run to get cold water or letting a hose run while washing a car. Another problem is that leaks are often ignored because the wasted water costs the customer nothing.

Water Quality

Customers will generally use less water if it has an unpleasant taste, odor, or color. They may also use less if it is hard or highly mineralized.

Sewers

The availability of municipal sewer systems has also been found to increase water use. Homes with sewer service usually have more appliances and fixtures that use water because there is no worry about overloading a private septic system. The increase in water use due to sewer availability may be 50 to 100 percent.

Condition of the Water System

When water mains are installed, they are tested for water tightness, but as they get older, leakage in pipe joints, valves, and connections increases. Unless periodic checks are made to locate and repair leaks, the amount of water lost daily can equal or exceed the amount used by consumers. If the amount of lost water is significant, it can place an extra strain on the ability of the utility to furnish water to customers, and it also represents a major loss of revenue.

Water Conservation

The last two decades have seen a steady progression in the acceptance of water conservation in the United States. Water utilities in water-short areas have been leaders in conservation efforts. Where water is plentiful, there is generally less interest in conservation programs.

For utilities in areas of prolonged drought, the benefits of water use reduction are obvious. The benefits are less obvious in an area that has adequate rainfall, a relatively stable population, and a source with an adequate safe yield.

Water Conservation Benefits

The benefits of water conservation are as follows:

- reduced demand on the supply source
- energy savings
- reduction of wastewater flow
- reduced costs
- protection of the environment

A reduction in water use, whether from a surface water or groundwater source, can allow a greater population to be served by the same size source. It can also extend supplies during an emergency. This is particularly applicable to aquifers that are becoming depleted and surface sources that serve the increasing demands of growing communities.

A reduction in water use can provide energy savings for the water utility. Pumping costs can be reduced, as can power and chemicals required for all stages of treatment. Energy savings resulting from reduced wastewater flows include both pumping and treatment power costs. Reduced flow can substantially extend the capacity of overloaded sewers and sewage treatment facilities. In addition, when heated water is conserved, energy costs will be reduced in homes, businesses, and industry.

Reducing the demands on a stream or surface water impoundment can minimize the impact of water use on the plants and animals that depend on it. This aspect of water conservation is receiving increasing attention from environmental organizations and governmental agencies.

Potential Problems With Water Conservation

Although the benefits of water conservation clearly outweigh the disadvantages, there are some negative aspects that must be considered. These include

- loss of revenue for the utility
- possible delay in developing additional source capacity
- possible stimulation of water service growth
- difficulty in dealing with drought conditions

Loss of Revenue

When a utility implements a water conservation program, it will receive less income because of reduced water sales. There will be some offsetting savings due to lower energy and chemical costs, but the revenue loss can still be substantial. Often the only way of compensating for this loss is to raise water rates.

A water system manager must realize before starting a conservation program that revenue will decrease. He or she must take steps to ensure that adequate revenue will be available to meet obligations such as bond payments. If a utility must suddenly implement water conservation because of a water shortage, the unplanned loss of revenue may be quite serious because it is difficult to raise rates on short notice. In other words, it is better to anticipate a water shortage and plan for it than to wait until it happens.

Delay in Developing Additional Water Sources

When a water crisis is alleviated through conservation, the public will find it hard to understand why there is still a need for long-range water resource development projects. Lack of public support could then delay development of a new reservoir, well field, or other new water source that is needed. This delay can, in turn, increase the cost of the project, as well as worsen the water crisis if there is a drought year before the new facilities can be constructed.

Stimulation of System Growth

The water saved by a conservation program will allow more people to be served by the amount of water available. This can act as a stimulant to growth. As long as water still appears to be available for use, additional customers will want to connect to the system — which may not be in the best interests of the water system or community, at least until a more ample supply can be obtained.

Dealing With Drought Conditions

When a drought occurs after the system already has a conservation program in effect, the situation becomes very difficult. It is hard to explain why it is necessary to conserve even further. This emphasizes the need for water supply planning as an integral part of conservation planning. The utility must insist that looking to the future water supply needs of a community is as important as any other aspect of a conservation program.

Supply Management Techniques

Conservation begins at home. The utility should examine its own practices in properly managing the available supply of water in conjunction with any water conservation program. This process is called supply management. Some features of supply management are

- careful management of all resources
- analysis of water use data
- complete source and customer metering
- reduction of unaccounted-for water

Maintaining an adequate source of supply should be part of both a long-range (25-year) and short-range (5-year) plan. The plan should be in writing and be made available to all local interest groups for comment and review.

A database on water use should be developed, including, but not limited to, data on average daily production, peak demand, and geographical and seasonal water use patterns. The complexity of this database should grow with utility size so that any unexpected change can be analyzed for its impact on total water use.

All sources of supply and customer services should be metered. Complete metering is necessary for comparing the amount of water used by customers against the amount pumped to the system. The difference between the amount of water pumped to the system and the amount metered to customers (or otherwise accounted for) is called unaccounted-for water.

Unaccounted-for water is generally water that is wasted through leaks or unauthorized use. State utility commissions usually have formulas that can be used to calculate unaccounted-for water. American Water Works Association (AWWA) guidelines suggest corrective action if the total exceeds 20 percent, but water systems having a conservation program should strive for a much lower percentage.

If the amount of unaccounted-for water appears to be substantial, a systemwide leak survey should be conducted. In addition, leak awareness for field personnel can be promoted by a program of listening for leaks on hydrants, services, and meters.

Demand Management Techniques

In conjunction with the program to reduce water use on the supply end, the utility must develop and implement demand management programs to encourage the public to reduce water usage. Demand management consists of the following:

- public education
- distribution of water-saving devices
- management of water use demands
- modification of the water rate structure
- use of water-saving plumbing fixtures
- promotion of low–water-use gardening and agricultural practices
- promotion of water conservation by businesses and industries

Public Education

Educating the public is the first step in starting a water conservation program. If water rates are low or moderate, customers need considerable incentive to participate in a water conservation program. Printed materials and suggested programs are available from AWWA, state water organizations, state agencies, and the US Environmental Protection Agency (USEPA), as well as environmental and conservation organizations. A good way of reaching and informing the public about the conservation program is by offering speakers, literature, and videos to local schools, service clubs, and civic groups.

Customers can also be educated directly by the use of bill stuffers and conservation notes. Direct assistance in detecting leaks or developing water use audits should be offered to customers. Newspapers, radio, and television stations should also be supplied with public service announcements (PSAs) and press releases explaining the program.

Water-Saving Devices

Many water utilities offer water-saving kits to residential customers. These kits reduce water use by altering the flow of water through regular plumbing fixtures. A typical kit includes faucet aerators to decrease faucet flow, a low-flow showerhead or flow restrictor, and a toilet tank dam to reduce the amount of water used by each flush. Some utilities have even offered their customers rebates for replacing older-model toilets with low-flow models. The cost of these kits can be fully paid by the customer, partially paid by the customer, or fully funded by the utility. State regulatory agencies have mandated the use of these kits in many areas and have stipulated that the cost be shared by the customer and utility.

Managing Water Use Demands

Some sections of a water system may experience water shortage as a result of inadequate distribution system capacity. If this occurs, the utility may gain some relief by educating the public about the problem and encouraging them to use water during off-peak hours. A common summer water use restriction can mandate that water for lawn watering or car washing be used only on alternating days. These restrictions can be either mandatory or voluntary.

Rate Structure Modification

Rate structure changes are gaining in popularity as a way to encourage water conservation. There are a variety of choices, and information on them can be obtained from AWWA and public utility regulatory agencies. A first step can be to implement a flat rate per unit volume of water. The customer must then pay the same rate for all metered water, rather than have the rate decrease with increasing use, as is normally the case. If it becomes necessary to exert greater pressure on customers to conserve water, some utilities have implemented rates that increase as greater amounts of water are used.

Water-Saving Plumbing Fixtures

Some states that have continuing water shortages have passed legislation, as part of the state plumbing code, mandating use of water-saving fixtures. Some jurisdictions have also required fixtures to be labeled with information on water use efficiency.

Low-Water-Use Gardening Practices

The use of landscaping requiring very low water use is called Xeriscaping (Figure 5-4). It is being encouraged for home and business landscaping in semiarid and arid areas as a means of conserving irrigation

FIGURE 5-4
Typical
water-efficient
landscaping

Source: Doug Hanford, Hanford Company — Landscape Design; John Nelson, North Marin
Water District; Ali Davidson, Sonoma County Water Agency.

water. There are excellent publications on this subject. (See Suggested Supplementary Readings at the end of this chapter.) The use of drought-tolerant landscaping also has value in areas where there are seasonal shortages of water.

Business and Industry Conservation

Utilities can often achieve great savings in water use by working with business and industry. Increasing water rates is a very strong motivation for these groups. However, in many cases, little thought has been given to how water is used or wasted in a manufacturing plant, and all that is needed is some careful consideration of how use can be reduced. In some instances, installing smaller, more sensitive meters in businesses and industries can generate considerable revenue by more accurately recording low flows, and can serve as a stimulus to water conservation.

Droughts

The most serious threat to the ability of a water utility to meet the demands of its customers is a drought. The actions required to meet a drought emergency are to locate an additional supply, to reduce demands, or both.

Each water system should have an emergency response plan addressing how water will be provided under all conceivable disasters. Although a drought can be very serious, immediate response is not usually necessary. There is usually a period of weeks or even months when it becomes obvious that source supplies are dwindling. Therefore, finding additional sources of water and reducing water use should not be done during an acute crisis.

Emergency procedures to reduce water use are similar to routine water conservation, but the actions required will be more severe when the situation is critical. Some communities have gone as far as instituting fines for unauthorized uses, such as washing a car or watering the lawn, during an acute water shortage.

Water Rights

From the beginning of recorded history, people have recognized the value of water and argued over its ownership. In the United States, the issue of water rights is one that is growing in importance and intensity. There are many arguments over who owns a source of water and who gets to use it. The arguments may be between individuals, between communities, between states, between states and the federal government, and between countries.

Allocation of Surface Water

The two basic systems for the allocation of water from surface lakes and streams in the United States are

- the riparian doctrine, commonly used in the water-rich eastern part of the country
- the appropriation doctrine, used in the more arid western states

Figure 5-5 shows which type of water rights system is used in each state, and Figure 5-6 shows which states require surface water permits.

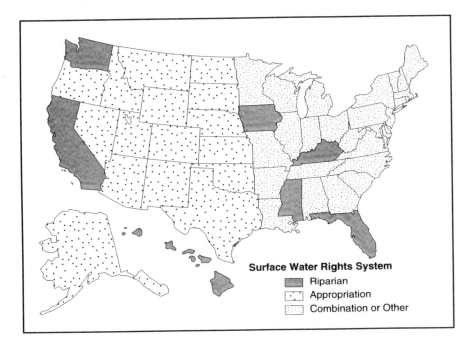

**Figure 5-5
Surface water
rights system**

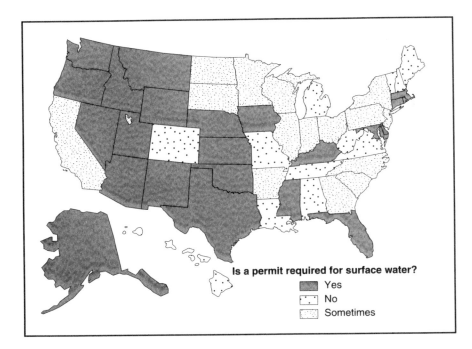

**Figure 5-6
Surface water
permit
requirement**

Riparian Doctrine

The riparian doctrine is sometimes called the "rule of reasonable sharing." The basic theory in all states that follow this doctrine is that there should be equal sharing among those who own the land abutting a body of water. However, there are substantial differences in how the law is applied.

The basic principle is sharing, rather than a right to a specific amount of water or a certain rate of flow. In general, the riparian property owner can use as much water as desired for any reasonable purpose, as long as this use does not interfere with the reasonable use by other riparians.

Thus, the presence of a water shortage means that all riparians suffer equally and each must accept a reduction in supply. The rights of an owner of land abutting a water source are not affected by how long the land has been owned, or by how much water the owner has used in the past.

Riparian rights represent a common-law doctrine. This means that they are a part of a large body of civil law that, for the most part, was made by judges in court decisions in individual cases, rather than statutory law enacted by a legislative body.

Because the possession of rights arises from ownership of land abutting a water body, all of the owners are treated as equals by the courts. Conflicts are then usually resolved by the court, which determines what is reasonable use. This is not clearly defined and is not an absolute. Many factors and values are considered, including

- the purpose of the use
- its suitability to the water body
- its economic or social value
- the amount of harm caused by the use
- the amount of harm avoided by changing either party's use
- the protection of existing values

The growth of water use in riparian-doctrine states has led to more frequent conflicts created by water shortages. A number of these states have tried to remedy weaknesses in the doctrine by enacting statutes that require permits to be obtained from a state agency. These permits vary widely in the restrictions they apply and the requirements they place on the riparian.

This move to the issuance of permits brings the states closer to the appropriation doctrine. The permit systems are criticized by some and considered unnecessary by others. They have not undergone serious challenge in the courts, so their eventual success is not guaranteed. The trend toward permit systems, however, is continuing.

Appropriation Doctrine

The appropriation doctrine concept started with the forty-niners in the gold fields of California. When there was too little water for all the mining operations, the principle applied was "first in time, first in right." The principle was given legal recognition by the courts and later made into law by western state legislatures. Currently, the most important water use in the West is for irrigation, so to a great degree, the appropriation doctrine is irrigation law.

The doctrine rests on two basic principles: priority use and beneficial use.

Priority use. When stream flow is less than demand, use is prioritized based on who has been using the water for the longest period of time. The most recent uses are discontinued in order to provide water for the earlier uses. There is no sharing of suffering, which is a hallmark of the riparian doctrine.

Beneficial use. Beneficial use is, in many ways, quite the opposite of riparian doctrine. There is no water right based on land ownership alone. Instead, water rights are acquired by proving that water is being used beneficially. A user who has been appropriated water continues to be entitled to it only when it can be used beneficially. Waste is therefore theoretically prohibited, and any available water beyond the amount that can be used by one appropriator is available to others. In addition, nonuse of the water for a long period may result in loss of the water right by forfeiture or abandonment.

Beneficial use is seldom defined precisely. There are two different but related aspects: the type of use (such as irrigation, mining, or municipal use) and efficiency. Most use is judged beneficial, and the majority of court cases have been about inefficient or wasteful uses.

Legal complications. When we consider disputes over real estate, rights are reasonably clear. People respect boundaries in most cases, and lawsuits are a suitable means of settling the few disputes that do occur. Water rights are different. Water is a shared resource, making water rights highly interdependent and relative. Boundaries are not obvious and are constantly changing.

With the appropriation doctrine, the assignment of a right at any particular time depends on stream flow, the number of prior existing rights, and the amount of water needed by the person or entity with the prior rights. For many western streams, the exercise of rights might have to be adjusted daily in times of heavy use, such as in summer. Irrigation uses, for example, make quick adjustments imperative, and the nature of the stream and river systems can complicate the issue.

Allocation of Groundwater

The following four water rights systems are used for groundwater:

- absolute ownership
- reasonable use
- correlative rights
- appropriation–permit systems

Absolute Ownership

The absolute ownership system is based on the principle that the owner of the land owns everything beneath that land to the center of the earth. A better understanding of the nature of groundwater makes it impossible to apply this ownership literally. For all practical purposes, the system is a rule of capture. The landowner can use all the water that can be "captured" from beneath the owner's land.

There are almost no restrictions applied to this rule. The water can be used for the owner's purposes both on and off the land, and it can be sold to others. There is no liability if pumping reduces a neighbor's supply or even dries up the neighbor's well. Some interpretations even say that water can be wasted as long as it is not done maliciously.

Reasonable Use

As is true with the rule of absolute ownership, the rule of reasonable use holds that groundwater rights are part of land ownership. The owner has the right to pump and use groundwater. However, if this use interferes with a neighbor's use, it may continue only if it is reasonable. Conversely, an owner is liable if unreasonable use causes harm to others. This amounts to a *qualified right* rather than an *absolute right* to use groundwater.

It is obvious that the determination of what is reasonable is the important factor in the rule. Reasonable use is much easier to determine for groundwater than for surface water under the riparian doctrine. In general, any nonwasteful use of water for a purpose associated with the land from which the water is drawn (an overlying use) is reasonable. Conversely, any use off the property may be considered unreasonable if it interferes with the use of the groundwater sources by others.

The overlying use criterion, however, does not resolve disputes if all parties to the dispute are using the water for overlying purposes. In this case, all uses are reasonable, all can continue, and "reasonable" use becomes the rule of capture — in other words, the owner with the biggest pump gets the water. On the other hand, because a nonoverlying use is considered unreasonable if it interferes with an overlying use, such uses are forbidden no matter how beneficial they might be.

Correlative Rights

The rule of correlative rights holds that the overlying use rule is not absolute, but is related to the rights of other overlying users. This rule is used when there is not enough water to satisfy all overlying uses. In this case the rule requires sharing. This rule has been applied by prorating the available water supply according to the size of the overlying land parcel.

Appropriation–Permit System

The appropriation–permit system is sometimes called a groundwater appropriation system, but it is in reality more of a permit system. The most important issue here is the rule of priority (water rights are based on who has used the water for the longest period of time). Priority is all-important in the appropriation doctrine as applied to surface water because the variation in stream flows requires frequent adjustment of withdrawals. However, groundwater in the aquifer is stored to a certain extent, and this storage evens out the supply so that frequent adjustment is not necessary. This means that priority, as applied to groundwater, mainly involves limiting the number of permits to prevent overuse of the aquifer.

The permit system thus amounts to groundwater management and administrative regulation. By placing limitations on pumping rates, well field placement, and well construction standards, and by refusing permits when necessary, for example, the administrating agency manages groundwater. Permit systems vary widely. The only generalization that can be made is that permits can be more relaxed if adequate water is available and more stringent if it is not. Figure 5-7 indicates which states have groundwater permit requirements.

Summary

There is a great variety in how rights are interpreted and how water rights are managed in US states and territories. The evolution of water resources management is more complete in the West, where water is frequently in short supply, than in the East, where supplies have been more plentiful. Management beyond the basic principles described here varies

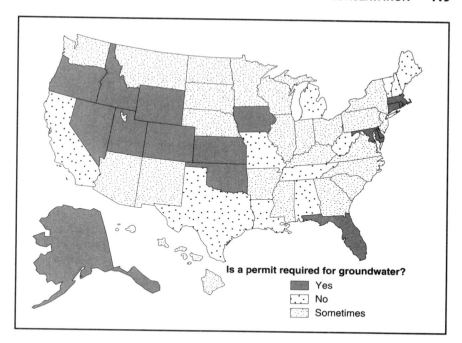

Is a permit required for groundwater?
- ■ Yes
- ⬚ No
- ⬚ Sometimes

Figure 5-7
Groundwater
permit
requirement

from state to state. The water utility operator should contact the state agency responsible for water resource management and obtain copies of applicable regulations.

Selected Supplementary Readings

Back to Basics Guide to Water Conservation. 1991. Denver, Colo.: American Water Works Association.

Ball, Ken. 1990. *Xeriscape Programs for Water Utilities.* Denver, Colo.: American Water Works Association.

Bowen, P.T., J.F. Harp, J.W. Baxter, and R.D. Shull. 1993. *Residential Water Use Patterns.* Denver, Colo.: American Water Works Association Research Foundation and American Water Works Association.

Driscoll, F.G. 1986. *Groundwater and Wells.* St Paul, Minn.: Johnson Filtration Systems Inc.

Drought Management Planning. 1992. Denver, Colo.: American Water Works Association.

Evaluating Urban Water Conservation Programs: A Procedures Manual. 1993. Denver, Colo.: American Water Works Association.

Federal Environmental Law Constraints on Drinking Water Supply Development. 1992. Denver, Colo.: American Water Works Association.

Gilbert, J.B., W.J. Bishop, and J.A. Weber. 1990. Reducing Water Demand During Drought Years. *Jour. AWWA,* 82(5):34.

Helping Businesses Manage Water Use. 1993. Denver, Colo.: American Water Works Association.

Kollar, K.L., and P. MacAuley. 1980. Water Requirements for Industrial Development. *Jour. AWWA,* 72(2):1.

Macy, P.P., and W.O. Maddaus. 1989. Cost–Benefit Analysis of Conservation Programs. *Jour. AWWA,* 81(3):43.

Maddaus, W.O. 1987. *Water Conservation.* Denver, Colo.: American Water Works Association.

Solley, W.B., R.R. Pierce, and H.A. Perlman. 1993. *Estimated Use of Water in the United States in 1990.* US Geological Survey Circular 1081. Washington, D.C.: US Government Printing Office.

The *Water Conservation Manager's Guide to Residential Retrofit.* 1993. Denver, Colo.: American Water Works Association.

van der Luden, F., F.L. Troise, and D.K. Todd. 2nd ed. 1990. *The Water Encyclopedia.* Chelsea, Mich.: Lewis Publishers.

Wright, Kenneth. 1990. *Water Rights in the Fifty States and Territories.* Denver, Colo.: American Water Works Association.

CHAPTER 6

Water Quality

It is important that all public water systems serve water of the best possible quality to their customers. If the quality is poor, customers will generally reduce consumption, and if it is very bad, they will look elsewhere for water.

If customers do obtain drinking water from another source, health officials fear that the substitute could be unsafe, especially if it comes from an unprotected roadside spring. The substitute, bottled water, for example, might also be very expensive. It is undesirable for the customer to lose confidence in the water utility and the water industry. The utility may then find it difficult to obtain approval for rate increases and hard to rally public support for improvements such as developing new water sources, upgrading treatment, and improving the distribution system.

A loss of confidence in the utility on the part of the public is difficult to overcome. It can be overcome, however, if the utility, the regulator, environmental groups, and the media work together to reassure the public of the safety of the product. Water utility operators must take the lead in this cooperative effort by ensuring that the goals of potability and palatability are met. The actions of the utility must be open to the public, and the customers should be kept informed through bill stuffers and public service announcements. The utility may also sponsor special information programs in conjunction with publicizing National Drinking Water Week.

Water Quality Characteristics

Water quality characteristics fall into the following four broad categories:

- physical
- chemical

- biological
- radiological

The quality of water is based to a great extent on the concentrations of the impurities that are present. The smallest unit of water, the water molecule (H_2O), contains two atoms of hydrogen (H) and one atom of oxygen (O). Anything present in the water other than water molecules may be considered an impurity. In the natural environment, there is no such thing as pure water. Even rain and snow contain impurities.

This is even true of distilled water, which often has traces of chemicals that are not completely removed in the distillation process. In fact, industries that require very pure water, such as the electronic chip industry, use elaborate treatment facilities to produce very costly water that is almost completely free of impurities. Because water is capable of dissolving trace amounts of almost anything, highly treated water must be used immediately, rather than being stored.

Water Contaminants

It is now common practice for the impurities in water to be called contaminants. Many contaminants are harmless, and some are even beneficial. Examples of beneficial contaminants include fluoride and calcium in normal concentrations.

Measuring Contaminant Levels

The concentration of a particular contaminant in water is usually very small and is measured in milligrams per liter (mg/L). This means that the quantity of contaminant in a standard volume (a liter) of water is measured by weight (in milligrams). A concentration of 1 mg/L of magnesium in water would be about the same as 0.00013 ounces of magnesium in each gallon of water.

In water, concentrations expressed as milligrams per liter, within the range of 0–2,000 mg/L, are roughly equivalent to concentrations expressed as parts per million (ppm). For example, 12 mg/L of calcium in water is roughly the same as 12 ppm calcium in water.

Mineral Concentrations

The mineral content of any water consists of individual chemicals, such as calcium, magnesium, sodium, iron, and manganese. The examples given in Table 6-1 provide a general idea of the relative magnitude of mineral concentrations in various types of water.

TABLE 6-1 Typical surface water mineral concentrations

Source of Water	Total Dissolved Minerals, mg/L
Distilled	<1
Rain	10
Lake Tahoe	70
Suwannee River	150
Lake Michigan	170
Missouri River	360
Pecos River	2,600
Ocean	35,000
Brine well	125,000
Dead Sea	250,000

In general, river water has a dissolved-minerals concentration of less than 500 mg/L, although some rivers may have concentrations of 2,000 mg/L or more. Mineral concentrations in groundwater can vary from several hundred to more than 10,000 mg/L.

Physical Characteristics

The physical properties of water that are important in water treatment include

- temperature
- turbidity
- color
- tastes and odors

Temperature

Water temperature is quite important in water treatment. For example, chemicals used for treatment are dissolved more easily in warm water, and suspended particles will also settle out more quickly. However, warmer temperatures encourage the growth of various forms of plant life in water.

From the consumer's standpoint, cold water is preferred because it tastes better. When industries use water for cooling and process-manufacturing, they generally prefer cold water.

The temperature of water in lakes and reservoirs generally varies with depth. The heavier cold water will normally be at the bottom, with warmer water near the surface. However, as air temperatures become colder in the fall, the surface water cools and eventually becomes slightly heavier than the water below. At that point, the lake "turns over" as the cold surface water sinks to the bottom. The turbulence created by a lake turnover can disperse decomposed bottom deposits throughout the reservoir. This can create a very unpleasant odor in the water and can cause quality and treatment problems for several days or weeks.

The reverse of this process usually occurs in the spring. By knowing the water temperature at various levels in a reservoir, the operator is more likely to anticipate turnover and select the level of withdrawal that will produce the best-quality water. Methods of preventing lake stratification are discussed in another book in this series, *Water Treatment*.

Temperature is measured on either of two scales: Fahrenheit or Celsius. The Fahrenheit scale is primarily used in the United States. The Celsius scale is used by most other countries and in scientific work. Relationships between the two scales are as follows:

Fahrenheit (F):	Freezing point = 32°F
	Boiling point = 212°F
	°F = $\frac{9}{5}$ (°C) + 32
Celsius (C):	Freezing point = 0°C
	Boiling point = 100°C
	°C = $\frac{5}{9}$ (°F − 32)

Turbidity

The cloudiness caused by suspended matter in water is called turbidity. The suspended matter consists of finely divided solids that are larger than molecules but generally not distinguishable by the naked eye. The solids are of organic origin and include clay, silt, and asbestos fibers, as well as living and dead organic matter such as algae, bacteria, and other microorganisms.

When a beam of light passes through turbid water, some of the light reflects off the suspended particles. Measurements of the intensity of this light indicate the amount of turbidity present. Turbidity is expressed in terms of nephelometric turbidity units (ntu).

Although the presence of turbidity is an aesthetic concern in drinking water, it is even more of a health concern. The particulates may themselves be toxic or may have adsorbed toxic material. There is a particular concern that harmful microorganisms adsorbed to the particles will be shielded against disinfectants and will thus remain viable.

Because of the natural filtering effect of soils, groundwater turbidity is often near zero. Surface water turbidities vary widely from less than 1 ntu to as high as 200 ntu or more.

Color

Color in raw water is primarily a problem for surface waters. It usually indicates the presence of decomposed organic material such as leaves, roots, or plant remains. It may also be due either to a high concentration of inorganic chemicals or to domestic or industrial wastewater. Color in water can sometimes indicate pollution that might be harmful.

Color is measured by comparing the color of a sample with the color of a standard chemical solution. The units of measure are color units (cu). A color of less than 15 cu usually passes unnoticed, whereas a color of 100 cu has the appearance of light tea. Highly colored water is objectionable for most industrial uses, and from an aesthetic standpoint, it is unsuitable for drinking water. Figure 6-1 shows a colorimeter used to measure color.

Tastes and Odors

Tastes and odors in water can be caused by a wide variety of materials including algae, decaying organic matter, industrial and domestic wastes, minerals, and dissolved gases. Highly mineralized waters have a saline, medicinal, or metallic taste. On the other hand, small amounts of minerals and dissolved gases can give water a pleasant taste. Tastes and odors become more noticeable as the water temperature increases.

Although the senses of taste and smell are closely related, the sense of smell is far more discriminating. Therefore, in water treatment, usually only odor is measured. The intensity of odor is measured by a person who smells a series of diluted water samples. The person begins the test by smelling a sample of specially prepared, odor-free water and continues by smelling samples that contain increasing concentrations of the water being tested. This continues until an odor is detected. The minimum odor that can be detected is called the threshold odor. The particular dilution of that sample is called the threshold odor number (TON). The TON test is conducted on raw water to determine the degree of treatment required. The test is also regularly conducted on finished water to ensure that adequate treatment is being provided.

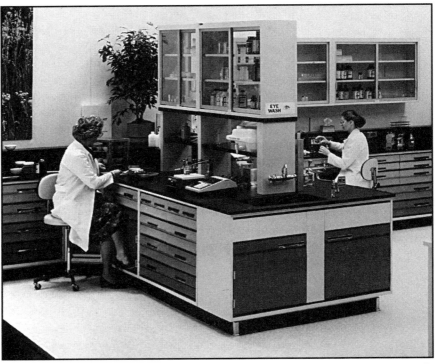

Courtesy of Fisher Scientific

FIGURE 6-1
Colorimeter
used in treatment
plant lab

Chemical Characteristics

Chemical characteristics of water fall into two categories: inorganic and organic. Organic chemicals have a basically carbon-based structure, whereas inorganic chemicals do not.

Inorganic characteristics include

- pH
- hardness
- dissolved oxygen
- dissolved solids
- electrical conductivity

pH

The term pH is used to express the acidic or alkaline condition of a solution. The pH scale runs from 0 to 14, with 7 being neutral. A pH less than 7 indicates an acidic water, and a pH greater than 7 indicates a basic, or alkaline, water. The normal pH range of surface water is 6.5 to 8.5; the pH of groundwater generally ranges from 6.0 to 8.5. A typical pH meter used to determine the pH of water is shown in operation in Figure 6-2.

The pH of water can have a marked effect on treatment plant equipment and processes. At a pH value less than 7, water has a greater tendency to corrode the equipment and other materials that it contacts (Figure 6-3). At a pH value greater than 7, water has a tendency to deposit scale, which is particularly noticeable in pipelines and hot-water home appliances (Figure 6-4).

The pH of water also has an effect on treatment processes such as coagulation and disinfection. If raw water is very far from neutral, the pH usually needs to be adjusted to about 7.0 or slightly higher before the water is pumped to the distribution system.

Hardness

A "hard" water is any water that contains significant amounts of calcium and magnesium. Hard water can be a problem in water supply and treatment because, like water with a high pH, it can cause scale to form in pipes and meters.

FIGURE 6-2 pH testing in the field

Courtesy of Fisher Scientific

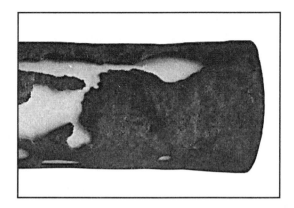

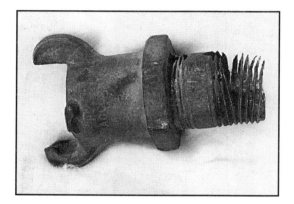

FIGURE 6-3
Corroded pipe
and appurtenance

If water is hot, as in boilers and hot-water lines, the scale forms much faster. One millimeter (0.04 in.) of deposited scale can increase the cost of heating hot water by more than 10 percent. If water from a public water system is very hard, many customers install individual home water softeners.

Hard water may have an objectionable taste, and it requires more soap when used for washing. A standard laboratory test for hardness determines the amount of both calcium and magnesium in the water, but for convenience, the results are expressed in milligrams per liter as calcium carbonate ($CaCO_3$). Table 6-2 summarizes typical degrees of hardness and the corresponding concentrations.

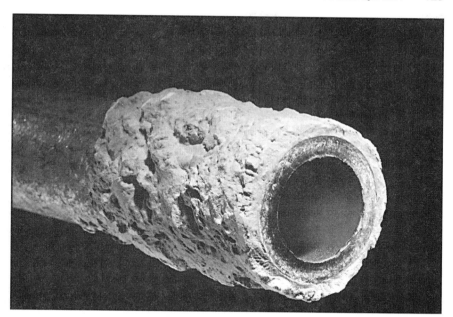

FIGURE 6-4 Pipe scaling caused by high pH values

Courtesy of Johnson Controls, Inc.

TABLE 6-2 **Hardness concentrations and typical corresponding designations**

Common Designation	Hardness, *mg/L as CaCO₃*
Soft	0–60
Moderately hard	61–120
Hard	121–180
Very hard	Greater than 180

Dissolved Oxygen

One of the more common dissolved gases in water is oxygen. Dissolved oxygen (DO) is absolutely vital for the support of fish and other aquatic life.

Oxygen can become dissolved in water in three ways.

- The oxygen enters the water directly from the air (through natural aeration).
- Oxygen is introduced into the water by algae (through photosynthesis).

- Oxygen is introduced by mechanical equipment during treatment, such as mechanical aeration or diffused aeration.

Dissolved oxygen is almost always present to some extent in natural waters. A warm-water lake will usually contain about 5 mg/L of DO; a cold-water lake may contain 7 mg/L or more. However, if algae are present in the water, the DO level will vary greatly. During sunlight hours, algae produce oxygen, so the DO level rises. During the night, algae use the oxygen, causing the DO level to drop.

Although DO is a healthy and necessary characteristic of water, it is also an important cause of corrosion. Water containing high levels of DO will attack metallic surfaces that it contacts, such as pipes, meters, pumps, and boilers. In general, the higher the DO concentration, the more rapid the corrosion will be.

DO can be measured either chemically or electrically. The measured concentration of DO is reported in milligrams per liter. A typical DO meter is shown in Figure 6-5.

FIGURE 6-5
Dissolved oxygen
meter in use

Courtesy of Fisher Scientific

Dissolved Solids

Because of its remarkable dissolving properties, water dissolves minerals from the soil and rock materials that it contacts. Table 6-3 demonstrates the wide variation in water quality that occurs in water from four different sources.

Water can at times contain many toxic dissolved minerals, including arsenic, barium, cadmium, chromium, lead, mercury, selenium, and silver. Other dissolved minerals, such as dissolved iron and manganese, can cause aesthetic problems, for example, turning water brown or black when the chemicals are oxidized.

The total quantity of dissolved solids in water is a general indicator of its acceptability for drinking, agricultural, and industrial uses. Water with a high concentration of dissolved solids can create tastes, odors, hardness, corrosion, and scaling problems. In addition, high concentrations of dissolved solids have a laxative effect on most people. For these reasons, a dissolved solids limit of 500 mg/L is recommended for drinking water.

Industrial users usually require water with a relatively low concentration of dissolved solids in order to prevent corrosion and boiler scale. If the water

TABLE 6-3   Chemical quality comparison of four water sources

Chemical	Concentrations of Water Contaminants, *mg/L*			
	River*	Well†	Canal‡	Saline Lake§
Silica (SiO_2)	5.4	41	6.6	11
Iron (Fe)	0.11	0.04	0.11	0.10
Calcium (Ca)	9.6	50	83	2.9
Magnesium (Mg)	2.4	4.8	6.7	9.5
Sodium (Na)	4.2	10	12	8,690
Potassium (K)	1.1	5.1	1.2	138
Carbonate (CO_3)	0	0	0	3,010
Bicarbonate (HCO_3)	26	172	263	3,600
Sulfate (SO_4)	12	8.0	5.4	10,500
Chloride (CI)	5.0	5.0	20	668
Fluoride (F)	0.1	0.4	0.2	—
Nitrate (NO_3)	3.2	20	1.3	5.8
Total dissolved solids	64	250	310	25,000

*Stream in Connecticut
†Logan County, Colorado
‡Drainage from the Everglades in Florida
§North-central North Dakota

is to be incorporated into a product, the level must be quite low. On the other hand, certain irrigated crops can tolerate mineralized water at a concentration as high as 2,000 mg/L.

Dissolved solids are measured by filtering a known volume of sample and then evaporating the filtered water to dryness. The residue that remains is weighed, and the results are recorded in milligrams per liter as filterable residue, commonly called total dissolved solids (TDS). This test is called the filterable residue test.

Electrical Conductivity

A common way to obtain a quick estimate of the concentration of dissolved solids in water is to measure the electrical conductivity (EC) of the water. Theoretically, water free of any contaminants will not conduct electricity. However, most natural substances dissolved in water allow water to conduct electricity — the greater the concentration, the greater the electrical conductivity.

The instrument used to determine EC actually measures the electrical resistance of the water between two electrodes spaced a known distance apart. The conductivity instrument readout is in micromhos per centimeter at 25°C, abbreviated μmhos/cm at 25°C. As a general rule, every 10 units of EC represents 6 to 7 mg/L of dissolved solids. Therefore, an EC of 1,000 suggests a dissolved solids concentration of about 600–700 mg/L. It should be noted that the relationship between EC and dissolved solids is somewhat different for every type of water.

An EC test takes only a few minutes, whereas the filterable residue test may take two hours to complete. The test is temperature sensitive, so it must always be conducted with the sample at 25°C. Electrical conductivity is measured by a conductivity meter like the one shown in Figure 6-6.

Organic Characteristics

Organics are another category of chemical characteristics. To date, more than 1,000 organic chemicals have been identified in drinking water supplies. Organic chemicals have four general sources.

- Plant decomposition yields materials such as tannins, lignins, and fulvic and humic materials.
- Municipal and industrial wastewater discharges may add natural and synthetic organics.
- Agricultural runoff yields synthetic organics such as pesticides and herbicides.

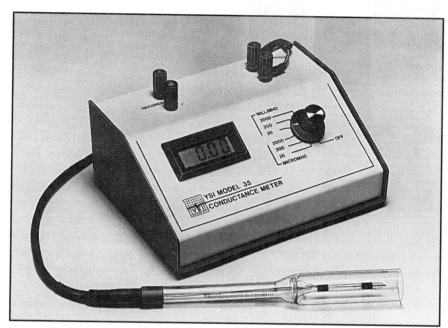

FIGURE 6-6
Typical electrical
conductivity meter

Photo provided by YSI Incorporated

- Water treatment operations can produce complex organics, such as trihalomethanes, during the disinfection process.

Most natural organic chemicals found in water are harmless, but some synthetic organic chemicals (SOCs) are considered to be carcinogenic (cancer causing) or cause other adverse health effects at relatively low concentrations. In addition, some organic materials that are harmless in their natural state can react during disinfection processes to produce suspected cancer-causing compounds. Organic chemicals can also cause aesthetic problems such as tastes, odors, and color in water.

Biological Characteristics

Biological characteristics correspond to the plants and animals, both dead and alive, that are found in water. They include viruses, bacteria, and other forms of aquatic life, as well as animal and plant contaminants.

Algae

Algae are an example of aquatic plants that can seriously affect the quality of surface water. Some algae cause taste-and-odor problems; others clog sand filters or produce slime growth on equipment, tanks, and reservoir walls. They are present to some extent in almost all surface water. Control of algae and other aquatic plants in lakes and reservoirs is discussed in *Water Treatment,* another book in this series.

Bacteria and Viruses

Various types of bacteria can clog well screens, produce discolored water in the distribution system, or cause taste-and-odor problems.

Of particular concern are bacteria and viruses that can cause disease in humans. These organisms do not ordinarily live in the water; they enter it through human or animal contact with the source water.

Microscopic Animals

There are some disease-causing microscopic animals that are not naturally found in water but can survive for a period of time after being deposited by humans or animals. An example is *Giardia lamblia,* a protozoan that causes an intestinal disease in humans.

Radiological Characteristics

Radionuclides can occur in water supplies either from natural sources or as a result of human activities. Naturally occurring radionuclides include radium 226, radium 228, and radon. None of these is usually present in significant amounts in surface waters, but they do frequently occur in groundwater. Radon is a gas most often present in granite formations, where it flows through the fractures in the rock to an aquifer.

Radioactivity produced by human activities can also enter surface water supplies from a variety of sources. Fallout from nuclear weapons testing was at one time the most significant source of radioactivity in surface water, but the level gradually diminished after testing was stopped. Other potential sources include waste from medical, scientific, and industrial users of radionuclides; wastewater discharge from nuclear power plants; and discharges from the mining of radioactive materials.

Factors Influencing Source Water Quality

Source water is affected primarily by the land around it, or in the case of groundwater, by the land over it. Land use practices can have a great effect on both surface water and groundwater, and the control of those practices is an important step in source water protection.

Surface Water Quality

Surface Runoff

After rain strikes the earth and becomes surface runoff, the water begins to dissolve materials that it contacts, and it takes on the characteristics of the land it passes over. Water flowing from remote areas of a watershed will not contain contaminants caused by human activities, but water passing over inhabited areas may contain many additional chemical and microbiological contaminants picked up in the flow across farmland and urban areas.

Agricultural runoff can contribute animal wastes, herbicides, pesticides, and fertilizer. Urban areas contribute a wide variety of organics and inorganics from homes, streets, pet wastes, storm water overflows, sewage outfalls, and other sources.

Point and Nonpoint Sources

All waste contributions to a surface water source can be classified as either point or nonpoint sources. These terms are now widely used in environmental and pollution control regulations. They are important to the water supply operator because of their impact on water sources.

In simplest terms, a point source is a waste flow that comes out of a pipe. Under state and federal pollution control regulations, the quality of wastewater from point sources is regulated to prevent pollution of lakes and streams.

Wastes that flow over the ground into a lake or stream are termed nonpoint sources. These sources of water contamination are much harder to control because the source is often indefinite.

There are many opportunities for water contamination, and some of them may be due to sources hundreds of miles away. The factors that influence source water quality may be categorized as natural or human factors. Natural factors include

- climate
- watershed characteristics
- geology
- microbial growth
- fire
- saltwater intrusion
- density (thermal stratification)

Human factors associated with point sources include

- wastewater discharges
- industrial discharges
- hazardous-waste facilities
- mine drainage
- spills and releases

Human factors associated with nonpoint sources include

- agricultural runoff
- livestock
- urban runoff
- land development
- landfills
- erosion
- atmospheric deposition
- recreational activities

Water From Reservoirs

Once runoff water enters a lake or storage reservoir, there can be other changes that take place — some beneficial and some not. As water enters a reservoir, suspended material in the moving water begins to settle when the water becomes quiescent or slow moving. While water is being held in a reservoir, bacterial action and natural oxidation can act on soluble contaminants and suspended particles to make them less likely to cause color, taste, and odor problems; sometimes the material will be converted into particles that will settle.

If a reservoir is large, it can absorb and dilute the shock loads of particularly turbid water received from streams following a storm. It may even be able to absorb and dilute a quantity of poor-quality water or a small chemical spill brought in by a stream. Wind action and algae usually keep the dissolved oxygen in a reservoir at beneficial levels, which helps natural purification processes and supports aquatic life.

Most bacteria that are pathogens, as well as protozoa that cause disease, are not natural to fresh water. Although some bacteria can survive for a fairly long time after leaving a human or animal, they cannot multiply. Eventually, most organisms will become nonviable and noninfectious when held in a reservoir for a period of time.

Algae are plants that contain chlorophyll. This means they can grow in water bodies exposed to light. There are three general categories of algae: diatoms, green algae, and blue-green algae.

Diatoms and green algae are found in reservoirs that have low nutrient levels and they help keep the water oxygenated. If they "bloom," or grow in profusion, they can clog intake screens and filters. At times, they can also cause taste-and-odor problems.

Blue-green algae are more commonly found in nutrient-rich waters and can bloom profusely, especially when water temperatures are higher. Blue-green algae are notorious taste-and-odor producers and probably account for the bulk of such problems. Even though algae produce oxygen, the die-off of a sizable bloom can deplete the water of oxygen as bacterial decomposition of the dead algae proceeds.

All lakes progress through a natural process called eutrophication. In the early stages, plant life and fish are not produced. Then they move to a more advanced state of productivity, and then to a point of overproductivity. Eventually, sediments fill the lake, and it becomes swampy or marshy. The eutrophication process usually takes centuries. All lakes are in one of the three stages, or somewhere in between.

The artificial introduction of nutrients into a lake from human activities can speed the eutrophication process. Lakes subjected to heavy loads of unnatural nutrients will have more frequent and serious algae blooms, and they may become unusable as water sources. Stopping the nutrient inflow or applying special treatment, such as aeration or chemical treatment, can sometimes restore such a lake to a useful state.

Additional details on treatment of lakes and reservoirs for algae and water weed control can be found in another book in this series, *Water Treatment*.

Atmospheric Contaminants

Airborne contamination is a nonpoint source that is very important and has received much attention in recent years. It is usually referred to under the general term "acid rain."

The most common component of acid rain is sulfuric acid, which is generated in the atmosphere from the combustion by-products of coal used for power generation. It was believed for many years that building a high smokestack was the solution to the problem of coal combustion emissions. Local residents were not affected and the pollutants were carried "far away."

We now know that, by a process known as photochemical oxidation, sulfur dioxide in the air is converted to sulfuric acid. This acid can be carried hundreds of miles and then deposited as acid rain, acid snow, and even acid dust. Many lakes in the northeastern United States and Canada have had a definite change in the pH of the water in recent years due to acid rain. This has caused severe environmental damage to aquatic life as well as solubilizing metals such as aluminum. A secondary effect is the damage to forests, which serve to reduce erosion on watersheds.

Other sources of airborne contaminants are automobile exhaust, emissions from waste incinerators, volcanic eruptions, and almost any combustion process that does not include complete treatment of stack gases. The passage of strong clean-air legislation in 1992 and expected further strengthening of clean-air standards should help reduce the impact of atmospheric contamination.

Groundwater Quality

The quality of groundwater tends to be uniform throughout an aquifer. Changes in quality occur slowly. Groundwater is not as subject to direct pollution and contamination as surface water.

Aquifer Contamination

Because of the protective soil cover and the natural filtration provided by the aquifer material, the biological characteristics of groundwater are generally very good. Harmful bacteria do not penetrate very far into the soil, so wells more than 50 ft (15 m) deep are generally free of harmful organisms. Some exceptions to this are aquifers fed by large voids in fractured rock, or locations where there are holes (such as improperly abandoned wells) penetrating into the aquifer. Even in cases where harmful bacteria do reach the aquifer, the hostile environment does not allow them to remain viable very long.

Most cases of groundwater contamination are a result of either poor well construction or poor waste disposal practices on the land overlying the aquifer. The water in most aquifers moves quite slowly and is not subject to the purifying action of air or sunlight. This means that contamination of groundwater can be a very long-lasting problem. Well supplies that are found to be contaminated must be abandoned or the water treated before it is supplied to customers. In some cases, barrier wells can be used to intercept contaminated water that is moving in the aquifer toward a public water supply well.

Chemical Characteristics

Groundwater acquires its chemical characteristics in two ways: from the surface water that percolates into the aquifer and from the soil and rock that it contacts. Normal recharge for aquifers is the rain that falls on the aquifer recharge area. This water can be clean when the rain falls on land that is free from pollution, or not clean if the land is polluted. An example is agricultural land. Aquifer recharge from farmland water runoff can be contaminated with nitrates, herbicides, and pesticides.

Not all contamination of groundwater comes from surface contamination. A major problem in coastal areas is saltwater intrusion. Under natural conditions, the fresh water and salt water are in equilibrium. When the fresh water is removed from the aquifer faster than it is replenished, the salt water moves toward the wells. Under these conditions, wells may still be used, but the withdrawal must be limited to prevent further saltwater intrusion.

There are many other natural contaminants affecting the potability of groundwater. Some elements and compounds are beneficial or harmless in small amounts and become troublesome or harmful only when found in excessive quantities.

There are other contaminants that are harmful at very low levels, such as barium, arsenic, and various radionuclides. Gases that are undesirable in groundwater include hydrogen sulfide, methane, and radon. Common groundwater contaminants and their associated problems are listed in Table 6-4.

Chemical Contaminants

Probably the most serious contaminants found in public water supply wells are manufactured organic solvents. Over the years, chemicals have been carelessly disposed of on the ground, and they are now being found in many well supplies. Some of these chemicals are considered a health threat at concentrations far less than can be detected by taste or smell. The only way they can be detected is by relatively complicated and expensive tests. If chemical contamination of a well is detected, it is generally best to abandon the well and use another water source. If there is no alternative source of water available, the chemicals can usually be removed from the water by aeration or carbon adsorption processes.

One of the most common contaminants that makes groundwater unfit to drink is gasoline. Thousands of wells, mostly private individual home wells (but some public supplies, too), have been contaminated because of leaking

TABLE 6-4 Common groundwater quality problems

Constituent	Source	Type of Problem	USEPA Primary Drinking Water Standards*	USEPA Secondary Drinking Water Standards†
Inorganic Constituents				
Arsenic	Naturally occurring‡	Toxic	0.05 mg/L	
Fluoride	Naturally occurring	Stains teeth, can cause tooth damage at high levels	2.4 mg/L	
Hydrogen sulfide	Naturally occurring	Offensive odor, flammable, corrosive		0.05 mg/L
Iron	Naturally occurring	Stains plumbing fixtures and laundry, causes tastes		0.3 mg/L
Manganese	Naturally occurring	Discolors laundry and plumbing fixtures, causes tastes		0.05 mg/L
Nitrate	Fertilizer and fecal matter	Toxic to infants	10 mg/L (as nitrogen)	
Radioactivity	Naturally occurring	Cancer causing	Gross alpha activity (15 pCi/L) Radium 226 and 228 (5 pCi/L)	
Sodium	Naturally occurring	May contribute to high blood pressure		Being investigated

*State regulations may be more restrictive.
†Not enforceable at federal level.
‡The naturally occurring elements can also be present as a result of contamination by humans.

Source: **Basics of Well Construction.**

Table continued next page

TABLE 6-4 Common groundwater quality problems (continued)

Constituent	Source	Type of Problem	USEPA Primary Drinking Water Standards*	USEPA Secondary Drinking Water Standards†
Inorganic Constituents (continued)				
Sulfate	Naturally occurring‡	Laxative effect		250 mg/L
Total dissolved solids	Naturally occurring‡	Associated with tastes, scale formation, corrosion, and hardness		500 mg/L
Organic Constituents				
Pesticides and herbicides	Agricultural and industrial contamination	Many are toxic, cause tastes and odors	Several are regulated	
Solvents	Industrial contamination	Many are toxic, cause tastes and odors	Being developed	
Microbiological Constituents				
Disease-causing microorganisms	Fecal contamination	Cause variety of illnesses	Coliform bacteria are regulated as indicator organism	
Iron bacteria	Contamination from surface	Produce foul-smelling slimes, which plug well screens, pumps, and valves		
Sulfate-reducing bacteria	Contamination from surface	Produce foul-smelling and corrosive hydrogen sulfide		

*State regulations may be more restrictive.

†Not enforceable at federal level.

‡The naturally occurring elements can also be present as a result of contamination by humans.

Source: **Basics of Well Construction.**

underground gasoline tanks. Less widespread, but not uncommon, is contamination from underground fuel oil storage.

Public Health Significance of Water Quality

Water quality can have a significant effect on public health as a result of

- waterborne diseases
- inorganic chemical contaminants
- disinfectants
- organic chemical contaminants
- radioactive contaminants

Waterborne Diseases

In spite of modern technology and the careful operation of most water treatment plants, waterborne disease outbreaks continue to occur. Table 6-5 provides a summary of the reported waterborne disease outbreaks in the United States in recent years. In most cases, the treatment techniques needed to prevent waterborne diseases from entering the finished water are available, but some part of the treatment process fails or is not operated correctly.

Even though modern treatment will remove or inactivate known disease organisms to safe levels, it is still best if the source water is as free of contamination as possible. Table 6-6 lists the causes and health effects of common waterborne diseases. It should be noted that the source of the organism in the water is, in most cases, human feces. It is therefore obvious that preventing human feces from entering water sources is the first step in preventing a disease outbreak. Human waste can originate from a point source such as a sewage outfall, or from a nonpoint source such as the flow of waste over the ground from a failed septic or cesspool system.

The single most prevalent waterborne disease is giardiasis. Most outbreaks that are traceable to a water supply originate from unfiltered water systems, but there have also been outbreaks caused by filtered water systems that were not functioning properly. *Giardia lamblia*, the cause of the disease, is a protozoan that can infect both humans and animals. Two common carriers are beavers and muskrats, which by their very nature are commonly found in surface water bodies. The organism has a cyst form that can remain viable under adverse conditions in water for one to three months. The cysts are

TABLE 6-5 Etiology of waterborne outbreaks in US water systems, 1971 to 1985

Illness	Number of Outbreaks	Cases of Illness
Gastroenteritis, undefined	251	61,478
Giardiasis	92	24,365
Chemical poisoning	50	3,774
Shigellosis	33	5,783
Hepatitis A	23	737
Gastroenteritis, viral	20	6,524
Campylobacterosis	11	4,983
Salmonellosis	10	2,300
Typhoid	5	282
Yersiniosis	2	103
Gastroenteritis, toxigenic *E. coli*	1	1,000
Cryptosporidiosis	1	117
Cholera	1	17
Dermatitis	1	31
Amebiasis	1	4
Total	502	111,228

Source: Craun (1988).

quite resistant to chlorine and can pass through filters that are not operating efficiently.

Waterborne disease organisms are extremely difficult to detect directly. In spite of many advances in laboratory technology, only very large, well-equipped laboratories can perform the tests for specific organisms. However, there are a number of nonharmful organisms that also indicate the presence of animal or human feces. The coliform group has been found over the years to be the most suitable. This group is the basis for the most important biological test that is run on water, the coliform test. There are test standards for both source water and treated water from customers' taps. If laboratory results show coliform bacteria to be present, the water has been contaminated by human or animal wastes, and it is assumed that the water is also capable of containing other disease-causing organisms (pathogens).

Inorganic Chemical Contaminants

Inorganic contaminants are primarily the metals found in water, but there are some other compounds and materials that are included. Legal standards have been established for the allowable concentrations of most of

TABLE 6-6 Potential waterborne disease-causing organisms

Name of Organism or Group	Major Disease	Primary Sources
Bacteria		
Salmonella typhi	Typhoid fever	Human feces
Salmonella paratyphi	Paratyphoid fever	Human feces
Other salmonella	Salmonellosis	Human and animal feces
Shigella	Bacillary dysentery	Human feces
Vibrio cholerae	Cholera	Human feces
Enteropathogenic *E. coli*	Gastroenteritis	Human feces
Yersinia enterocolitica	Gastroenteritis	Human and animal feces
Campylobacter jejuni	Gastroenteritis	Human (and possibly animal) feces
Legionella pneumophila and related bacteria	Acute respiratory illness (legionellosis)	Thermally enriched waters
Mycobacterium tuberculosis	Tuberculosis	Human respiratory exudates
Other (atypical) mycobacteria	Pulmonary illness	Soil and water
Opportunistic bacteria	Variable	Natural waters
Enteric Viruses		
Enteroviruses		
Polioviruses	Poliomyelitis	Human feces
Coxsackieviruses A	Aseptic meningitis	Human feces
Coxsackieviruses B	Aseptic meningitis	Human feces
Echoviruses	Aseptic meningitis	Human feces
Other enteroviruses	Encephalitis	Human feces
Reoviruses	Mild upper respiratory and gastrointestinal illness	Human and animal feces
Rotaviruses	Gastroenteritis	Human feces
Adenoviruses	Upper respiratory and gastrointestinal illness	Human feces
Hepatitis A virus	Infectious hepatitis	Human feces
Norwalk and related gastrointestinal viruses	Gastroenteritis	Human feces

Source: Water Quality and Treatment *(1990).*

Table continued next page

TABLE 6-6   Potential waterborne disease-causing organisms (continued)

Name of Organism or Group	Major Disease	Primary Sources
Protozoa		
Acanthamoeba castellani	Amoebic meningo-encephalitis	Soil and water
Balantidium coli	Balantidosis (dysentery)	Human feces
Cryptosporidium	Cryptosporidiosis	Human and animal feces
Entamoeba histolytica	Amoebic dysentery	Human feces
Giardia lamblia	Giardiasis (gastroenteritis)	Human and animal feces
Naegleria fowleri	Primary amoebic meningoencephalitis	Soil and water
Algae (blue-green)		
Anabaena flos-aquae	Gastroenteritis	Natural waters
Microcystis aeruginosa	Gastroenteritis	Natural waters
Aphanizomenon flos-aquae	Gastroenteritis	Natural waters
Schizothrix calciola	Gastroenteritis	Natural waters

them, though other contaminants are still being considered. Some of them are highly toxic, and some are only mildly toxic. The list includes

Aluminum	Copper	Selenium
Ammonia	Cyanide	Silver
Antimony	Fluoride	Sodium
Arsenic	Lead	Strontium
Asbestos	Manganese	Sulfate
Barium	Mercury	Thallium
Beryllium	Molybdenum	Vanadium
Boron	Nickel	Zinc
Cadmium	Nitrate	
Chromium	Nitrite	

Disinfectants

Chlorine is an inorganic chemical contaminant, but it is generally listed along with other disinfectants. Although chlorine is now being considered as either a direct or indirect cause of some adverse health effects, it should be remembered that its use has saved thousands (if not millions) of lives since it was introduced as a water disinfectant. State and federal regulations are steadily moving toward requiring mandatory chlorination of all public water supplies.

Other disinfectants in common use are chlorine dioxide, ozone, and potassium permanganate. These chemicals have some advantages for use in drinking water disinfection, but they have the principal disadvantage of not providing a lasting disinfectant residual that will remain in the water for a period of time.

Other methods of disinfecting water include the use of iodine, bromine, silver, hydrogen peroxide, or ultraviolet light. However, these methods are rarely used for disinfection of public drinking water supplies.

Chlorine

All forms of chlorination, including gaseous chlorine, liquid chlorine, sodium hypochlorite, and calcium hypochlorite, produce hypochlorous acid in dilute water solution. For killing or inactivating pathogens, hypochlorous acid is the most effective of the various chlorine compounds that occur when water is chlorinated. The chlorine disinfectants are all strong chemicals and are very toxic to plants and animals in high concentrations. At the low concentrations used for water disinfection, they are considered to have very low toxicity.

Free chlorine reacts with some of the natural organic compounds present in raw water to form varying amounts of chemicals that fall into a group called trihalomethanes (THMs). The principal THM is chloroform. These compounds, as a group, are considered carcinogenic. Consequently, state and federal regulations limit the amount of THMs that may be present in water provided to the public. It has also been found that chlorine and other disinfectants react with contaminants in water to create many additional compounds, some of which can probably cause adverse health effects. These substances have been grouped under the general heading of disinfection by-products (DBPs). Disinfection by-products are still under study, but it appears certain that additional regulations will be imposed on disinfection practices in future years to minimize exposure of the public to certain DBPs.

Chloramine

Chloramine is a somewhat different disinfectant that is formed when chlorine reacts with ammonia in the water. Chloramine has some advantages in reducing tastes and odors in some water and producing a lower concentration of THMs, but it does not have the disinfectant power of free chlorine against bacteria. It is also considered relatively ineffective for inactivating *Giardia*.

Ozone

Ozone is a form of oxygen and a very powerful oxidant and disinfectant. It is widely used in Europe and is rapidly becoming more popular in the United States. It destroys organic matter and is very effective against bacteria, viruses, *Giardia*, and *Cryptosporidium*. It quickly disappears in water, however, and consequently does not have protective residual effect in the distribution system. When ozone is used as an initial disinfectant for water, it must be followed by the addition of chlorine to provide a disinfectant residual.

Ozone is produced by an electrical discharge in the presence of an oxygen-rich atmosphere. It must be produced on-site. Both the initial capital costs for equipment and the operating costs of using ozone are quite high in comparison with those for chlorine.

Organic Chemical Contaminants

Organic contaminants in drinking water come from the following three major sources:

- the breakdown of naturally occurring organic materials
- contamination from industrial, commercial, and domestic activities
- reactions that take place during water treatment and distribution

The first source is by far the largest. It consists of breakdown products from the decay of leaves and other plant pieces, aquatic animal death and decomposition, algae decomposition, and other natural aquatic life by-products.

The chemicals that enter water from industrial, commercial, and domestic activities include all those used in agriculture, found in sewage discharges, and washed from contaminated land or urban runoff. This group, by far, contains the greatest number of chemicals of concern — the solvents, pesticides, herbicides, and other organics used by the business community and the public. There are now regulations covering the use and disposal of many of these chemicals, and more of the chemicals will be controlled in future years.

One of the best-known groups of organic chemicals are the polychlorinated biphenyls (PCBs). Discharge of PCBs in industrial wastes has generated immense publicity and controversy. Laboratory tests indicate that very low concentrations of PCBs fed to animals can cause cancer. Another highly publicized chemical is dioxin, which appears to be a potent animal carcinogen. The principles, occurrence, and control of organic chemicals in drinking water are covered in detail in *Water Quality*, part of this series.

Radioactive Contaminants

Radioactivity is a very frightening subject for the general public. The terror of nuclear war is so much in the public consciousness that it has colored every mention of radioactivity or nuclear waste. In spite of this, the radioactivity found in water is almost always from natural sources. All of the radioactive contaminants are carcinogenic. The radionuclides of concern in drinking water are

- uranium
- radium 226
- radium 228
- radon
- thorium 230
- thorium 232

One unit for expressing radiation dosage is the rem. The rem can be related to lifetime cancer risks. Humans are subjected to some radiation every day from cosmic rays from space, radiation from the soil and rocks around us, and X rays. The average annual dose from all sources is about 200 millirems (mrem) per year. The US Environmental Protection Agency (USEPA) estimates that radioactivity in drinking water contributes about 0.1 to 3 percent of the annual dose received by an average person. This can vary greatly, however, and the radionuclide that is usually responsible is radon. It is estimated that between 500 and 4,000 public water supplies have radon at levels that create an equivalent dose of 100 mrem per year.

It is currently estimated that up to 30,000 water systems have radon at a level where some human risk can be calculated. However, by far the greatest public exposure to radon comes from soils that underlie house basements. The gas is released from the soil and can enter the living space through basement cracks if the basement is not sealed. The relative impact of waterborne radon and soil-released radon on public health is still a matter of some controversy.

Public Water Supply Regulations

Regulations for the operation and monitoring of water systems serving the public were developed by each state during the early 1900s, but the policies and degree of enforcement varied considerably among the states.

Each state's health department generally monitored water quality, enforced regulations, and provided technical assistance to water system operators.

Prior to 1974, federal involvement in regulating the provision of drinking water to the public consisted only of the development of drinking water standards by the US Public Health Service. These standards suggested bacterial and chemical limits for public water systems. Although the standards were not federally enforceable, they were incorporated into the regulations of most states.

The Safe Drinking Water Act

During the 1960s and early 1970s, scientists and public health experts discovered a previously unrecognized potential for harmful disease organisms and chemicals in drinking water. There was also increasing pressure by the public and legislators to create uniform national standards for drinking water to ensure that every public water supply in the country would meet minimum health standards.

Congress accordingly passed the original Safe Drinking Water Act (SDWA) in 1974. The SDWA directs the USEPA to establish standards and requirements necessary to protect the public from all known harmful contaminants in drinking water. It also asks each of the states to accept primary enforcement responsibility (primacy) for enforcing the federal requirements.

SDWA Provisions

There are many requirements relating to public water system operation, monitoring, and reporting in the SDWA, but a few of the more important points are as follows:

- The USEPA establishes primary drinking water standards for microbiological and chemical contaminants that may be found in drinking water and that could cause adverse health effects for humans. The maximum concentration that is allowable is called the maximum contaminant level (MCL). All states must make their regulations at least as stringent as the MCLs established by the USEPA. The primary standards are mandatory, and all public water systems in the United States must comply with them.

- The USEPA establishes secondary drinking water standards that set recommended MCLs for contaminants that are not a direct threat to public health but affect aesthetic qualities such as tastes, odors, color, and staining of fixtures. The secondary standards are strongly recommended but not enforceable.

- Public notification (PN) is required by the SDWA as the first step of enforcement against all public water systems that fail to comply with federal requirements. Systems that violate operating, monitoring, or reporting requirements, or that exceed an MCL, must inform the public of the problem and explain the public health significance of their violation.
- Federal enforcement, including stiff monetary fines, may be leveled against water systems that do not comply with the federal requirements.
- A public water system is defined as a system that supplies piped water for human consumption and has at least 15 service connections or 25 or more persons who are served by the system for 60 or more days a year.

Provisions of the SDWA and regulations enacted by the USEPA are explained in detail in *Water Quality*, which is also part of this series.

Classes of Public Water Systems

In implementing the requirements of the SDWA, the USEPA has divided public water systems into three categories based on the number and type of customers served.

- Community public water systems are those water utilities that serve full-time customers. Examples are municipal systems, rural water districts, and mobile home parks.
- Nontransient, noncommunity public water systems are entities having their own water supply serving an average of at least 25 persons who do not live at the location, but who use the water for over six months per year. Examples might include schools, factories, and office buildings.
- Transient, noncommunity public water systems are establishments having their own water system, but where people visit and use the water occasionally or for short periods of time. Examples might include parks, motels, campgrounds, restaurants, and churches.

Community and nontransient, noncommunity water systems are required to meet specific operating requirements and to perform relatively extensive monitoring of water quality. This is necessary because customers are using the water over a long period of time and could experience adverse health effects from continuous exposure to relatively low levels of contaminants. The requirements for transient, noncommunity systems are not as stringent because the customers use the water only occasionally.

Water systems having fewer than 25 customers and individual homes having their own source of water (such as a well) are referred to as private or nonpublic water systems. Nonpublic systems are not covered by the SDWA, but they are generally regulated by state and/or local jurisdictions.

State Drinking Water Programs

The intent of the SDWA is for each state to accept primary enforcement responsibility for the operation of the drinking water program within their state. To be given primacy, a state must implement all requirements, standards, and programs required by the USEPA. In return, the state receives a federal grant to supplement state funds to operate the public water supply program.

Most states also have programs in addition to those mandated by the USEPA. These include certification of water system operators, cross-connection control, and monitoring for contaminants not part of the federal requirements.

The major functions provided by state drinking water primacy agencies are as follows:

- *Monitoring and tracking*. The state staff must ensure that all systems perform the proper monitoring, use approved laboratories, and regularly submit reports on system operation.
- *Sanitary surveys*. A sanitary survey is an on-site inspection of a water system's facilities and operation. Surveys are performed by state staff or other qualified persons on a regular basis. Reports are prepared for the system owner or operator, and problems or deficiencies are identified.
- *Plan review*. The SDWA requires that states review and approve all plans for water system construction and major improvements.
- *Technical assistance*. One of the staff functions of the state program is to provide water system owners and operators with technical assistance when they have problems.
- *Laboratory services*. Chemical, radiological, and microbiological analyses of water quality must be performed by a certified laboratory. In some states, state laboratories perform all or most of the tests. Other states certify commercial laboratories that do the work for the water systems.

- *Enforcement.* The SDWA requires states to use enforcement actions when federal requirements are violated.

As the federal requirements are periodically changed and expanded, each state must, within a prescribed period of time, make similar changes in its regulations and begin implementation and enforcement.

Selected Supplementary Readings

Craun, G.F. 1988. Surface Water Supplies and Health. *Jour. AWWA,* 80:2.

Manual M7, Problem Organisms in Water: Identification and Treatment. 1995. Denver, Colo.: American Water Works Association.

Manual of Instruction for Water Treatment Plant Operators. 1975. Albany, N.Y.: New York State Department of Health.

Manual of Water Utility Operations. 8th ed. 1988. Austin, Texas: Texas Water Utilities Association.

Pontius, F.W. *SDWA Advisor: Regulatory Update Service.* Denver, Colo.: American Water Works Association.

VOCs and Unregulated Contaminants. 1990. Denver, Colo.: American Water Works Association.

Water Quality and Treatment. 4th ed. 1990. New York: McGraw-Hill and American Water Works Association (available from AWWA).

CHAPTER 7

Water Source Protection

The quality of source water is subject to threats both from nature and from human activities. It is the operator's responsibility to minimize harm from both of these threats. Surface water sources are more directly contaminated by human causes, but as has been found in recent years, groundwater sources can also be affected.

Fundamental Principles

Surface water can usually recover from pollution incidents much faster than groundwater. However, it is also more susceptible to the assaults of nature such as flooding, droughts, or algae blooms.

The changes in a groundwater source are much less frequent, but when they do occur, natural recovery is generally very slow. Some threats to a water source are relatively obvious, as in a gasoline storage tank being installed next to a well or a herd of cattle being brought in to graze on the shores of a small reservoir. Unfortunately, most threats are not this obvious, and many are difficult to identify, even with time and hard work.

Source Protection Area

The first step in source protection is to identify the area that needs protection and who has an interest in protecting it. All of the water system's customers should certainly be interested and should be kept informed. Farmers and business people who might be in the protection area have a financial stake in preventing pollution, so they should be involved. Municipal and county officials may have to act as an enforcement arm, so they should be involved, and public health and environmental protection agencies should also be kept informed. When all of the interested parties have been contacted, and, optimally, when a task force has been formed, the job can begin.

Aquifer Delineation

Groundwater comes from an aquifer that is recharged by percolating and infiltrating surface water. The only way to tell what surface water is recharging a well's aquifer is to define the land area over the portion of the aquifer used by the well. This land area is called the aquifer recharge area or the wellhead protection area (WHPA).

In the past, many states used a fixed radius around a well to determine the area to be protected. This was commonly 100 ft (30 m) for a private well and as little as 300 ft (90 m) for a public water supply well. Utilities were sometimes required to own the land within that circle or to at least have control over its use.

If a well is relatively deep and properly constructed, protection of a fixed radius around the well is probably adequate in most cases for protection from microbiological contaminants. However, it has been found to have little bearing on contamination from chemicals. Some chemicals have been found to travel up to thousands of feet within an aquifer. In other words, depending on local geological formations, the WHPA of a well could need to be a relatively small area or a very large area.

If the region is one that has already had its groundwater mapped by the US Geological Survey or a state geology department, information on the recharge area for a well may be readily available. Unless the area is one with very easily determined borders, it is best for a utility to hire a firm having experts in geology, hydrology, and mapping of aquifers.

Watershed Mapping

If a water utility uses a surface water source, the watershed boundary should be determined. The outside boundary of a watershed is the surface water divide between watersheds. All water falling as rain or snow flows toward gradually larger waterways. As illustrated in Figure 7-1, various types of existing maps can be used in watershed mapping and for identifying surface features.

US Geological Survey topographic maps are particularly useful for outlining the area. The ridges between drainage areas are easily identified. Water on one side of the ridge flows toward one drainage area, and water on the other side flows in another direction. Using lines to connect the ridges on

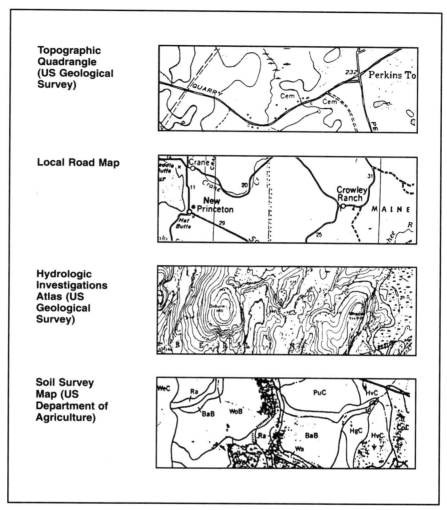

Topographic Quadrangle (US Geological Survey)

Local Road Map

Hydrologic Investigations Atlas (US Geological Survey)

Soil Survey Map (US Department of Agriculture)

FIGURE 7-1
Various types of maps that can be used for watershed mapping

Source: Protecting Local Ground-Water Supplies Through Wellhead Protection *(1991).*

the map will outline the area contributing water to the source used by a water system.

Once a watershed has been identified, the topographic maps can also be used to identify the general types of vegetation and activities that exist within the drainage area. Features such as forests, open land, roads, industrial activities, and houses that were present when the map was prepared are all shown.

Sources of Contamination

When the WHPA or watershed boundary for a water source has been determined, the next step is to develop a complete inventory of the potential contamination sources within the area. This does not have to be an expensive project, and it does not have to be done by water utility personnel. It could become a public service project. Local volunteers from scouting, senior citizens, service clubs, and volunteer fire department groups have proven very effective in doing the job.

Each volunteer or group is furnished with a map, or a portion of a map, and asked to mark potential contamination sources on it. A list of common sources, given in Table 7-1, should be furnished to the volunteers. Detailed information on locations of particular concern can be compiled by the volunteers, or followed up by more experienced personnel.

Local and area zoning maps and master plans should also be reviewed to identify possible future sources of contamination from industrial parks, waste disposal sites, and similar threatening developments.

Vandalism and Terrorism

Water utility personnel can be thankful that violence and violent acts have not been associated with water sources or water supply to any important extent. Even in those countries at war or suffering from civil unrest, water supplies have not been used as agents to carry toxic or disease-causing materials. However, the possibility of vandalism or terrorism aimed at a water system must be considered, and it should be part of any utility emergency plan.

If water was delivered to customers in small volumes to be used exclusively for drinking, it would be relatively easy for a terrorist to add enough dangerous material to cause human illness. However, the fact that drinking and cooking account for only a small portion of the water delivered to customers by a public water system is an important safety factor. It means that a rather large volume of poison would have to be introduced to the water supply to create a high enough concentration to cause sickness from drinking the water.

During World War II, many utilities took serious measures to safeguard water sources as well as treatment and pumping facilities. The threat to facilities is a much more believable scenario. Arson or use of explosives could easily prevent the delivery of water to at least a portion of the transmission and distribution system for most water utilities. Prevention of those acts is practical only in the face of a threat, but responding to the results of such acts

TABLE 7-1 Common sources of groundwater contamination

Category	Contaminant Source
Agricultural	Animal burial areas
	Animal feedlots
	Fertilizer storage or use
	Irrigation sites
	Manure-spreading areas or pits
	Pesticide storage or use
Commercial	Airports
	Auto repair shops
	Boat yards
	Construction areas
	Car washes
	Cemeteries
	Dry cleaners
	Gas stations
	Golf courses
	Jewelry or metal plating
	Laundromats
	Medical institutions
	Paint shops
	Photography establishments
	Railroad tracks and yards
	Research laboratories
	Scrap and junkyards
	Storage tanks
Industrial	Asphalt plants
	Chemical manufacture or storage
	Electronics manufacture
	Electroplaters
	Foundries or metal fabricators
	Machine or metalworking shops
	Mining and mine drainage
	Petroleum production or storage
	Pipelines
	Septic lagoons and sludge
	Storage tanks
	Toxic and hazardous spills
	Wells (operating or abandoned)
	Wood-preserving facilities

Table continued next page

TABLE 7-1   Common sources of groundwater contamination (continued)

Category	Contaminant Source
Residential	Fuel oil
	Furniture stripping or refinishing
	Household hazardous products
	Household lawns
	Septic systems or cesspools
	Sewer lines
	Swimming pools (chemicals)
Other	Hazardous-waste landfills
	Municipal incinerators
	Municipal landfills
	Municipal sewer lines
	Open burning sites
	Recycling or reduction facilities
	Road-deicing operations
	Road maintenance depots
	Stormwater drains or basins
	Transfer stations

should be considered a part of every utility's emergency plans. Fires and explosions are much more likely to happen as a result of human error or an accident than of malice.

Groundwater Protection

Groundwater protection can be instituted at federal, state, and local levels.

Federal-Level Aquifer Protection

Because a large number of people depend on it for their water needs, groundwater is an important resource that must be protected. Even though many serious groundwater contamination incidents have occurred, and new ones will continue to happen, the public and regulators are becoming more aware that protection is necessary. Figure 7-2 illustrates some of the many potential sources of groundwater contamination.

Sole-Source Aquifers

One mechanism for protecting an aquifer is to have it designated as a sole-source aquifer. Under provisions of the Safe Drinking Water Act

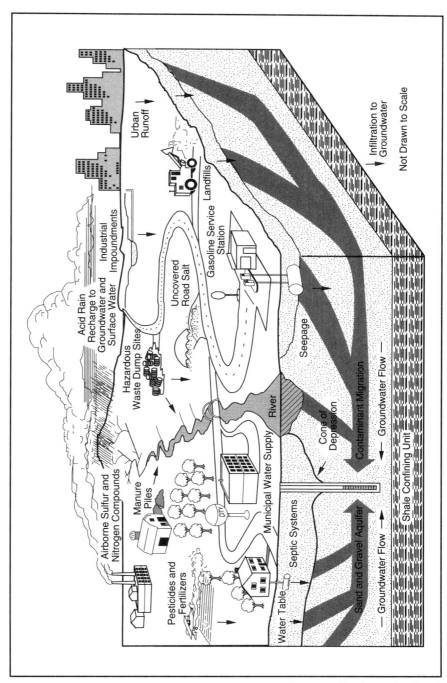

**FIGURE 7-2
Sources of
groundwater
contamination**

Source: Wellhead Protection *(1987).*

(SDWA), the US Environmental Protection Agency (USEPA) can, on its own initiative or as the result of a petition, designate an aquifer as a sole source for an area. After a determination is published, no federal money may be used or granted for any purpose that could result in the contamination of that aquifer.

Sole-source protection has been granted to aquifers as small as those located on rocky islands off the coast of Maine and as large as the 175-mi-long (282-km-long) Edwards Aquifer that supplies the San Antonio, Texas, area. The Edwards Aquifer was the first to receive sole-source designation under the SDWA.

Other Environmental Regulations

There are a number of other national laws that protect aquifers.

- The Resource Conservation and Recovery Act (RCRA) regulates the storage, transportation, treatment, and disposal of solid and hazardous wastes. It is principally designed to prevent contaminants from leaching into groundwater from municipal landfills, underground storage tanks, surface impoundments, and hazardous-waste disposal facilities.

- The Comprehensive Environmental Response, Compensation, and Liability Act (Superfund) authorizes the government to clean up contamination caused by chemical spills or hazardous-waste sites that could (or already do) pose threats to the environment. It also allows citizens to sue violators of the law and establishes "community right-to-know" programs.

- The Federal Insecticide, Fungicide, and Rodenticide Act (FIFRA) authorizes the USEPA to control the availability of pesticides that have the ability to leach into groundwater.

- The Toxic Substances Control Act (TSCA) authorizes the USEPA to control the manufacture, use, storage, distribution, or disposal of toxic chemicals that have the potential to leach into groundwater.

- The Clean Water Act authorizes the USEPA to make grants to the states for the development of groundwater protection strategies and authorizes a number of programs to prevent water pollution.

- The Safe Drinking Water Act also requires the USEPA to regulate underground disposal of wastes in deep wells.

Although these are all national laws, they are designed to be implemented by the states in cooperation with local government. A major reason for emphasizing local action is that groundwater protection generally involves specific decisions on land use. Most land use controls are implemented by local authorities based on state laws.

State-Level Aquifer Protection

All states have passed some type of legislation related to the protection of aquifers. These fall into a number of subject areas. Some states have minimal programs, whereas others have very comprehensive ones. State water utility associations are generally active in promoting protective legislation. General features of state programs are as follows:

- *Comprehensive planning* is a program that many states have required for local jurisdictions. Cities, towns, counties, and local districts have been mandated to develop plans that protect a state's groundwaters.

- *Groundwater classification* reviews all groundwater sources and categorizes them so that the amount of protection needed can be determined and applied.

- *Standard setting* evaluates groundwater to determine levels of contamination. It can also include required remedial action.

- *Land use management* is related to planning in that it determines whether uses can cause contamination, and, if so, then regulates those uses.

- *Funding proposals* are used to aid in contaminated water cleanup and in supplying alternative drinking water in crisis areas.

- *Agricultural chemical regulations* similar to federal laws are imposed so that these chemicals are used judiciously.

- *Underground storage regulations* are imposed to control possible leakage of hazardous material that is generally stored underground. They establish criteria for the registration, construction, installation, monitoring, repair, abandonment, and financial liability related to underground storage.

- *Water usage control* is exercised to allocate the use of water, with the goal of preventing saltwater intrusions or other overutilization problems.

State programs are the most valuable general aids to the water utility. All water utility managers should make sure there is adequate funding of state programs to protect all the waters of their state.

Wellhead Protection Programs

In addition to the resources available to the utility at the national and state levels, a wellhead protection program (WHPP) adds control at the local level. If at all possible, water utilities should lead in developing the WHPP that involves their groundwater source. The 1986 amendments to the SDWA require that each state develop a comprehensive wellhead protection plan, but that program largely depends on active participation at the local level.

A WHPP requires cooperation between the local civil authority and the water utility. A utility that leads this process is much more likely to be pleased with the outcome than one that has a WHPP imposed on it. The most important technical step is the determination of the recharge area as discussed previously. It is very likely that the utility will have to use its own budget for funding this effort.

Determining the Wellhead Protection Area

There are different ways of delineating the area that needs protection, and engineering or geological consultants have their own preferences as to which method does the best job. The state health, environmental, or geological agency that is responsible for WHPPs can give valuable suggestions. Figure 7-3 illustrates a wellhead protection area laid out on a topographic map. Numerical modeling is a technique that is highly successful in some situations. Based on parameters that include hydraulic conductivity, specific yield, and stream bed permeability, a computer model can be developed. This model can then be verified and modified after comparison with known events. The modeling stages might be as follows:

1. Select the study area and model grid.
2. Prepare data.
3. Calibrate the model.
4. Verify the model.
5. Check the model sensitivity.

Once a wellhead protection area has been determined, a determination should be made of the time of travel for percolating water to reach the well. The utility should be more concerned about possible pollution in a zone that has 30 days travel time than one that has 500 days.

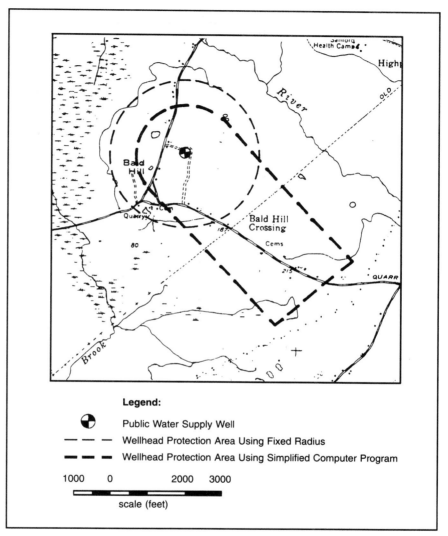

FIGURE 7-3
A wellhead
protection area
laid out on a
topographic map

Source: Protecting Local Ground-Water Supplies Through Wellhead Protection *(1991).*

Conducting a Contaminant Source Inventory

The second step in the WHPP is to conduct a contaminant source inventory. This can begin with a paper search in the phone book; chamber of commerce files; or police, fire, and other civil authority records. After the inventory is compiled, volunteers or trained personnel can conduct field work to obtain more specifics on particular contaminant sources.

Wellhead Protection Area Management

The WHPA and the contaminant sources and future development must then be managed. Local zoning can be of immense help in doing this. Citizens, authorities, and business and industrial leaders should be informed of the importance of protecting the water supply, and then asked to enact zoning laws that will control land use. Management of existing hazards can best be handled through personal contact and public education. The installation of monitoring wells may be necessary to ensure that major industries are not causing groundwater pollution. Some utilities have even convinced these industries to install the monitoring wells and pay for the sampling.

Contingency Plan

A WHPP is not complete until a contingency plan is prepared, detailing the steps to be taken if contamination ever occurs. The plan should provide for an alternative water supply for both short-term and long-term emergencies.

Continuing Review

Finally, the WHPP should be an active program. Built-in review periods with short intervals for minor review and longer periods for major review should be written into the plan.

Surface Water Protection

As an example of the comprehensive nature of surface water protection, the Massachusetts Watershed Protection plan provides the following statement, which applies to all watersheds, regardless of whether or not the water utility will be filtering the water:

> The protection of watersheds for ensuring the public's right to clean drinking water will require some creative and innovative approaches involving the cooperation of local, regional, and state levels of government. No agency alone can achieve the ambitious objectives required, because watersheds cover expansive areas, often encompassing several communities, and often include bordering states. Therefore, it will be in the public interest that all interested parties participate at each level of government and develop the cooperative mechanisms locally, regionally, and at the state level. The issue will be our greatest challenge and our greatest threat.

Watershed Resource Protection Plan

A watershed resource protection plan (WRPP) generally consists of the following elements:

- a description of the watershed
- identification of watershed characteristics and activities that are present and could be detrimental to water quality
- an assessment of the risks and how best to control detrimental activities and events
- how detrimental activities and events should be monitored
- determination of land ownership and what agreements are needed to control detrimental activities
- how best to manage and operate the program

The watershed description includes both a set of maps and a narrative description. The base map is a US Geological Survey (USGS) 1:25,000 map and has the following accompanying overlays of plastic sheeting, on which different information is drawn:

Base map	The watershed boundary
Overlay 1	Identification of points of water withdrawal, and an outline of aquifers that have the potential to yield greater than 100,000 gpd (380,000 L/d)
Overlay 2	Areas that have high erosion potential, and information on animal populations
Overlay 3	Identification of any state groundwater discharges having state or federal permits, all surface water discharges having National Pollution Discharge Elimination System (NPDES) permits, and solid waste facilities
Overlay 4	An outline of all sewered areas
Overlay 5	Location of all gasoline stations and petroleum storage facilities
Overlay 6	Identification of zoning areas
Overlay 7	Generalized land use, such as

- residential areas
- industrial areas
- commercial areas
- mixed urban areas
- croplands, orchards, golf courses

- animal-rearing areas
- other agriculture
- vacant or undeveloped land
- transportation facilities

Overlay 8 Hazardous-waste sites
Overlay 9 Water quality monitoring sites
Overlay 10 Land ownership

- open space owned or under the control of the water supplier, or state, county, or municipal government
- lands in conservation restrictions in perpetuity
- all lands owned or restricted by written agreements or deed restrictions with landowners

The benefit of transparent overlay maps as a watershed planning tool becomes evident once two or more of them have been superimposed on each other. They graphically show how the various potential problems coincide with each other. Once these maps have been used for watershed protection, they will become a regularly used tool. Many government agencies and utilities are now using a new technology known as geographic information systems (GIS) to display these various overlays on a computer monitor. For more information on this topic, refer to *Water Transmission and Distribution*, also part of this series.

It is unlikely that this project can be done without expert help and without spending a considerable amount of money. However, watershed planning is essential if a utility wishes to receive a waiver for the surface water filtration requirements of the SDWA. If the utility believes that the chances of a filtration waiver are good, then the cost of watershed planning will be small compared with the cost of constructing filtration facilities.

Sanitary Surveys

One of the most important control measures that any water utility can use to determine how well it is serving the public is the sanitary survey. A sanitary survey is an on-site review of the water source, facilities, equipment, operation, and maintenance of a public water system. It serves to evaluate the adequacy of all aspects of producing and distributing drinking water.

A sanitary survey of each water source is necessary for proper water source management. The utility may conduct it by following a standard form supplied by the state health authority. The form should be designed to identify all areas of concern within the supply system, including the entire watershed or aquifer recharge area. This portion should be completed by a qualified person, preferably a sanitary engineer or someone of equivalent training. The survey should determine the potential for contamination of the supply. This includes surveying all installations, activities, and contamination sources.

The survey, conducted periodically, should include water sampling, at places considered significant and representative, to establish a baseline for future comparison. Some portions of the system should be surveyed annually and others at less frequent intervals. Important changes in water quality found in annual sampling should be investigated. If the reason for the change cannot readily be determined, a complete survey should be performed.

Watershed Control Programs

Managing watersheds to maintain or enhance water quality is a particular necessity for utilities that wish to avoid filtration. It should, however, be a goal for all utilities using surface water. Obvious water quality hazards including sewage outfalls or toxic waste and landfill drainage must be eliminated by water utility managers. In fact, continuing activity on a watershed, though it may appear to be causing little or no damage, can be more dangerous in the long run. Table 7-2 shows some of the program elements for selected utility source management programs and illustrates the extensive watershed control program practiced by some water utilities.

Complete Control

The ideal situation from the standpoint of water source protection is a completely fenced, utility-owned watershed. This is usually not possible, but both Seattle, Wash., and Portland, Ore., do have this type of controlled supply.

Utilities that do not own an appreciable percentage of their watershed have to develop different protection strategies that depend more on the cooperation of the public, the local civil authorities, and the state. An example is the Portland, Maine, Water District, which serves a relatively small population but has a 450-mi^2 (1,166-km^2) watershed that includes portions of 21 communities (Figure 7-4).

The district makes a distinction between water that flows directly to Sebago Lake, and water that reaches the lake after passing through other lakes and ponds. The distinction is made because the latter water receives

TABLE 7-2 Elements of source water management programs for selected surface systems with highly protected managed watersheds

	Seattle, Wash.	Portland, Ore.	New York City, N.Y.	Boston, Mass.	Bridgeport Hydraulic Company, Conn.	Hackensack Water Company, N.J.
Type of supply	Reservoir, river	Reservoir, river	Reservoir	Reservoir, river	Reservoir	Reservoir
Safe yield, mgd (ML/d)	170 (643)	108 (409)	1,290 (4,883)	300 (1,136)	66 (250)	125 (473)
Filtered	No	No	No	No	Partial	Yes
Watershed area, mi^2 (km^2)	162 (420)	106 (275)	1,928 (4,994)	474 (1,228)	96 (249)	113 (293)
Program elements						
Watershed owned or controlled, mi^2 (km^2)	156 (404)	106 (275)	138(+) (357)*	180 (466)	26 (67)	10 (26)
%	100	100	7(+)	38	27	9
In situ treatment						
Aeration	–	–	–	–	X	–
Algicide or herbicide	–	–	X	X	X	X
Wildlife control	X	Partial	–	Partial	Partial	–
Forest management	X	X	X	X	X	X
Emergency response	X	X	X	X	X	X
Sanitary survey	X	–	X	X	X	X
Source monitoring	X	X	X	X	X	X
Watershed inspection	X	X	X	X	X	X
Security patrol	X	X	X	X	X	X
Fencing	–	Partial	–	Partial	–	X
Public education	X	X	–	X	X	X

*Approximately one-half of watershed dedicated to forest preserves.

Source: Water Quality and Treatment (1992).

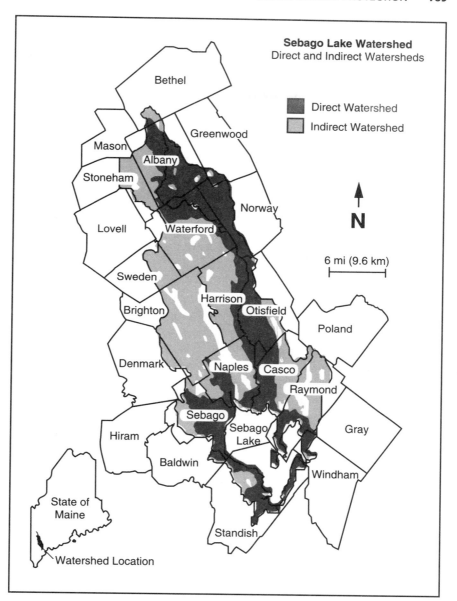

FIGURE 7-4
Sebago Lake
watershed,
Portland, Maine,
Water District,
which includes
portions of 21
communities

Source: Portland Water District, Portland, Maine.

some beneficial treatment in reducing phosphorus and other compounds. Portland's definitions for this distinction are as follows:

- *Direct watershed* consists of all land that drains directly to a lake without passing through another lake or pond.
- *Indirect watershed* of a lake consists of all watersheds of upstream lakes and ponds.

The Portland, Maine, source protection program calls on portions of 20 federal, state, and local laws. In addition, 10 to 15 sets of rules and regulations that carry the force of law are implemented when necessary. In most cases, when the need arises to invoke a state law, the appropriate state agency will take the lead in approaching the violator. The same is true for local problems — if the utility's watershed inspector finds a malfunctioning septic system, the local plumbing inspector or code enforcement officer is contacted to see that corrections are made.

Local Support

Very few utilities have police power, and thus, it becomes imperative for a utility to receive support from the authorities who do have the power. The best way to ensure this support is to have the backing of the public. Most citizens are aware of the environmental threats to water bodies and are willing to support the enforcement through corrective action. If it is "their" lake, water supply, or livelihood at stake, the public will readily help the utility.

As a result, public education is a vital part of a utility's source protection program. Operators should be available to talk to civil authorities serving on planning and zoning boards, to service clubs and other social groups, and especially to schools. Facilities should be maintained so that tours are always welcome and that the visitors will always receive a positive impression.

Members of the news media can be a valuable ally if they are convinced that watershed protection is necessary. It is the task of the utility to seek out newspaper, radio, and television sources so that an alliance can be forged. If the utility does not have someone on the staff capable of doing this, a public relations firm should be hired to do the job, at least for long enough to establish a cooperative relationship.

Watershed Protective Districts

Special watershed protective districts are another mechanism that has proven very effective in protecting water sources and watersheds. They are usually created when the geographical area to be protected is quite large and there are diverse interests desiring environmental protection.

One of the best known of these entities is the one that was organized to protect Lake Tahoe. This national treasure was being threatened by development and had the unusual feature of having its shoreline in two states (California and Nevada). Because of this, the protective legislation had to be passed by Congress.

In Maine, the Saco River Corridor Commission was founded by landowners for their own protection. However, the utility that uses the river as a water supply also benefits from the controls that are in place. The river extends a fairly long distance, which would be beyond the capabilities of the utility to protect if it had to act alone. However, the landowners have the power of eminent domain and the ability to level assessments.

Protective districts can be very difficult to get started because their scope can be quite broad and some interests do not see them as beneficial. In many cases, a water source has to be damaged or severely threatened before diverse groups perceive the threat and are willing to cooperate.

Even though these districts can help a utility by giving added protection to the watershed and water body, it may result in less control by the utility. Important decisions related to the watershed or water body, which were once made by the utility, will now be decided by a group comprised of a board of directors, of which the utility is only a one-vote member.

Land Management

In addition to the social and political aspects of watershed management, the actual practices of landowners or land users must be managed.

Forest Management

Forests are a familiar feature on watersheds in much of the country, and many utilities manage the forests as a source of income. Timber harvesting can be extremely damaging to the watershed and to water quality. Consequently, it is imperative that the tree cutter be environmentally responsible and that the operation be conducted according to the best timbering practices.

If the common practices presently in use do not adequately protect the water source, a consulting forester or someone familiar with best forest practice should be hired. An example of going beyond ordinary practices is a requirement that horses be used for tree-cutting work because motor-powered machinery is too damaging to the forest floor.

Much the same is true of home or commercial land development. If large areas are stripped of vegetation, extreme care must be exercised to make sure that erosion of the bare land does not damage the source water. Holding ponds and sediment basins that are constructed to minimize damage should be left in place after the project is completed so that the benefits they offer can continue.

Management of Agricultural Practices

Agricultural practices can be environmentally responsible or just the opposite. Farmers are usually in touch with conservation districts or extension agents (government officials who serve as consultants to agriculture), and the utility can enlist the aid of these experts to act as a liaison with the agricultural community. These experts can assist by encouraging the best practices, discouraging bad ones, and advising against new ventures that will potentially be harmful.

Management in Urban Areas

Homeowners close to a water body should be informed of the harmful effects of lawn and gardening practices that are not conducted properly. The use of nitrogen and phosphorous fertilizers to turn lawns green can also turn lakes green with algae.

Management of Transportation Facilities

If highways or railroads run through or close to water bodies, the emergency response agencies, namely police and fire departments, should know the proper procedure for dealing with spills. Flushing the contents of a serious spill off the highway to prevent fire or traffic disruption could put a water system out of business. For example, if a fire department uses water to extinguish a fire in an agricultural supply warehouse, they will probably flush pesticides and other chemicals into a nearby stream. They can usually contain the fire just as well with misting and allow it to burn itself out.

Recreational Use of Lakes and Reservoirs

Recreational use of water supply reservoirs or lakes has long been a subject of controversy. Studies have shown that controlled recreational use has minimal impact. However, water system managers generally have an aversion to the public being given free rein on water bodies that are used as drinking water sources because of the liability if anything goes wrong.

Because of population pressures from urban areas and the desirability of outdoor activities (especially those involving water), it is difficult to deny the public some use of watershed land. The goal then becomes one of managing recreational uses so that they do not threaten water quality.

Levels of Recreational Use

The following descriptions of recreational use range from a very low level of human activity to the very highest level. Many water utilities do not have the luxury of returning a watershed or water body to a low level of human contact if full recreational use is presently taking place.

The least impact on water quality occurs when recreation can be limited to bike and hiking paths, nature walks, and picnic areas. Even with these activities, it is necessary to provide adequate, clean, attractive toilet facilities and a plentiful supply of waste disposal containers that are serviced on a regular basis.

The above activities, plus fishing from shores and boats, form the next least-harmful recreational activity level. Boats should have no motors or small-horsepower motors. Launch sites should be as far from intake structures as possible, and restricted zones should be designated around intakes and marked with buoys.

The next step is to have less restriction on fishing and to allow boats with greater horsepower for nonfishing recreation. Speeds should be controlled to prevent erosion from wakes and to promote safety. Again, restricted areas should be enforced, and waste tanks on boats should be sealed so that wastes must be pumped out at an approved facility. Onshore sanitary facilities for people and boat holding-tank flushing stations should be provided. Fuel for boats should be kept as secure and free from spillage as possible.

Bodily contact with the water is the last step in this progression. Public swimming areas should be kept as far as possible from intake structures. Good onshore sanitary facilities should also be provided.

In all of these different scenarios, water quality monitoring and sanitary inspection are a necessity. It is obvious that as human contact increases, the more stringent the monitoring requirements become. If serious bacterial degradation occurs, regulators should be asked to help restrict activities.

Land Erosion

Erosion of watershed land adjacent to the shore seems to be the most important threat to water quality. It can result from footpaths and bike paths, but it can be much more serious with motor vehicles. While automobiles and trucks can be a source of erosion, the most serious threat comes from motorbikes, motorcycles, and all-terrain vehicles. The riders of these vehicles get the most enjoyment when they are able to use them off the beaten path. The greatest fun comes in climbing steep banks, which also leads to the most erosion and, if rain occurs, the greatest transfer of soil to the water body. Use of these vehicles should be restricted to areas where soil cannot be transported to the water body.

Recreational Fees

Another concept that has generally been accepted is that the public is more likely to follow the rules if they are charged for the use. Fees can also make the management of recreation on watersheds and water bodies less costly and thus more acceptable to taxpayers or customers. Care has to be taken, however, not to exclude those less able to pay.

Other Concerns

Many threats to water quality on the watershed and in the water body have not been listed here. Vigilance on the part of the utility, a spirit of cooperation with the public and civil authorities, and a well-operated treatment facility with well-trained, conscientious operators will result in a quality product being supplied to the water customer despite intensive recreational use of the watershed and water source.

The discussion of recreation applies only to raw-water sources. Recreation on finished-water reservoirs is universally condemned by the American Water Works Association and state health authorities and should be completely prohibited.

Filtration Waiver Requirements

A well-defined watershed control plan is mandatory for a utility to receive a waiver for filtration as required by the Surface Water Treatment Rule. Although filtration waivers are part of the national SDWA, the states that have assumed primacy have considerable flexibility in determining how the provisions will be enforced. The following information related to avoiding surface water filtration is part of the Massachusetts Watershed Protection Plan. Other states have similar plans. The plan's policy goals are as follows:

1. Provide guidance to water suppliers for preparing and submitting a report that describes a watershed control plan that will meet the USEPA's SDWA requirements to avoid

mandatory filtration and to serve as a watershed protection plan.

2. Provide specific and consistent guidance for the protection, preservation, and improvement of raw-water quality of surface water supplies. (These sources are open to the atmosphere and subject to surface runoff, or they may be groundwaters that originate from a surface water supply.)

3. Provide a comprehensive approach to identify and control existing and potential sources of pollution of surface water supplies.

4. Link land use activities to water quality concerns, and ensure that a consistent set of standards and safeguards are applied for reducing vulnerability to contamination from inappropriate land uses.

5. Encourage cooperation among the water supplier, landowners, and local authorities sharing a resource area to set up agreements to control discharges to the environment and to respond to emergencies.

6. Provide for an ongoing monitoring program that will be used by the Department of Environmental Protection to determine the effectiveness of the watershed controls on raw-water quality.

Selected Supplementary Readings

Citizens' Guide to Ground-Water Protection. 1990. Washington, D.C.: US Environmental Protection Agency, Office of Drinking Water.

Cooke, G.D., and R.E. Carlson. *Reservoir Management for Water Quality and THM Precursor Control.* 1989. Denver, Colo.: American Water Works Association Research Foundation and American Water Works Association.

Drinking Water Quality Enhancement Through Source Protection. 1977. Ann Arbor, Mich.: Ann Arbor Science.

Guide for Conducting Contaminant Source Inventories for Public Drinking Water Supplies. 1991. Washington, D.C.: US Environmental Protection Agency, Office of Drinking Water.

Local Financing for Wellhead Protection. 1989. Washington, D.C.: US Environmental Protection Agency, Office of Drinking Water.

Massachusetts Watershed Protection Plan. Boston, Mass.: Department of Environmental Protection.

Protecting Local Ground-Water Supplies Through Wellhead Protection. 1991. Washington, D.C.: US Environmental Protection Agency, Office of Drinking Water.

Robbins, R.W., J.L. Glicker, D.M. Bloem, and B.M. Niss. 1991. *Effective Watershed Management for Surface Water Supplies.* Denver, Colo.: American Water Works Association Research Foundation and American Water Works Association.

Stewart, J.C. 1990. *Drinking Water Hazards*. Hiram, Ohio: Envirographics.

Symons, J.M., M.E. Whitworth, P.B. Bedient, and J.G. Haughton. 1989. Managing an Urban Watershed. *Jour. AWWA*, 81(8):30.

Water Quality and Treatment. 4th ed. 1990. New York: McGraw-Hill and American Water Works Association (available from AWWA).

Wellhead Protection — A Decision Makers' Guide. 1987. Washington, D.C.: US Environmental Protection Agency, Office of Groundwater Protection.

GLOSSARY

absolute ownership A water rights term referring to water that is completely owned by one person.

absolute right A water right that cannot be abridged.

acid rain Rain or snow that has a low pH because of atmospheric contamination.

acidic water Water having a pH less than 7.0.

adsorption A physical process in which molecules adhere to a substance because of electrical charges.

advanced wastewater treatment The treatment of sewage beyond the ordinary level of treatment.

aeration The process of bringing water and air into close contact to remove or modify constituents in the water.

aesthetic Of or pertaining to the senses.

air-stripping Removal of a substance from water by means of air.

algae Primitive plants (one- or many-celled) that usually live in water and are capable of obtaining their food by photosynthesis.

alkaline water Water having a pH greater than 7.0. Also called *basic water*.

alluvial Referring to a type of soil, mostly sand and gravel, deposited by flowing water.

anaerobic Occurring in the absence of air or free oxygen.

annual average daily flow The average of the daily flows for a 12-month period. It may be determined by means of dividing the total volume of water for the year by 365, the number of days in the year.

annular space The space between the outside of a well casing and the drilled hole.

appropriation doctrine A water rights doctrine in which the first user has right to water before subsequent users.

appropriation–permit system A water use system in which permits to use water are regulated so that overdraft cannot occur.

appropriative rights Water rights acquired by means of diverting and putting the water to beneficial use following procedures established by state statutes or courts.

aquatic life All forms of animal and plant life that live in water.

aqueduct A conduit, usually of considerable size, used to convey water.

aquifer A porous, water-bearing geologic formation. Generally restricted to materials capable of yielding an appreciable supply of water. Also called *groundwater aquifer*.

aquifer recharge area The land above an aquifer that contributes water to it.

artesian aquifer An aquifer in which the water is confined by both an upper and a lower impermeable layer.

artesian well A well in which water pressure forces water up through a hole in the upper confining, or impermeable, layer of an artesian aquifer. In a flowing artesian well, the water will rise to the ground surface and flow out onto the ground.

artificial wetland A wet area created by a reduction of natural drainage in which biological action takes place.

atom The basic structural unit of matter; the smallest particle of an element that can combine chemically with similar particles of the same or other elements to form molecules of a compound.

average daily flow The sum of all daily flows for a specified time period, divided by the number of daily flows added.

bacteria A group of one-celled microscopic organisms that have no chlorophyll. Usually have spherical, rodlike, or curved shapes. Usually regarded as plants.

basic water See *alkaline water*.

beneficial use A water rights term indicating that the water is being used for good purposes.

biofilm A layer of biological material that covers a surface.

bloom An undesirable large growth of algae or diatoms in a lake or water reservoir.

buffered Able to resist changes in pH.

caisson A large-diameter excavated chamber in which work can be done beneath the surface of the earth or water.

capillary action   The degree to which a material or object containing minute openings or passages, when immersed in a liquid, usually water, will draw the surface of the liquid above the water surface.

carcinogenic   Capable of causing cancer.

casing   (1) The enclosure surrounding a pump impeller, into which are machined the suction and discharge ports. (2) The metal pipe used to line the borehole of a well (also called *well casing*).

cement grout   A mixture of cement and water used to seal openings in well construction.

chemical oxidation   The use of a chemical, such as chlorine or ozone, to remove or change some contaminant in water.

chemical precipitation   The use of a chemical to cause some contaminant to become insoluble in water.

chlorophyll   The green matter in plants or algae.

coagulation   A treatment that causes particles in water to adhere to each other to form larger particles.

coliform group   A group of bacteria predominantly inhabiting the intestines of humans or animals, but also occasionally found elsewhere. Presence of the bacteria in water is used as an indication of fecal contamination (contamination by human or animal waste).

color   A physical characteristic describing the appearance of water that is other than clear. Different from *turbidity,* which is the cloudiness of water.

color units (cu)   The unit of measure used to express the color of a water sample.

commercial water uses   Use of water in business activities such as motels, shopping centers, gas stations, and laundries.

common-law doctrine   Laws that are a part of a large body of civil law that, for the most part, was made by judges in court decisions (rather than being enacted by a legislative body).

compaction   A process that renders soil more dense and often reduces water transfer.

comprehensive planning   A process that results in widespread community planning.

computer model   A means of defining a system using computer programs.

condensation The process by which a substance changes from the gaseous form to a liquid or solid form. Water that falls as precipitation from the atmosphere has condensed from the vapor (gaseous) state to rain or snow. Dew and frost are also forms of condensation on the surface of the earth or vegetation.

cone of depression The cone-shaped depression in the groundwater level around a well during pumping.

confining bed A layer of material, typically consolidated rock or clay, that has very low permeability and restricts the movement of groundwater into or out of adjacent aquifers.

consecutive systems Two or more water systems that take water from a single providing system.

conservation district A governmental entity that acts to protect soil and water.

contaminant source inventory A record of the activities on a watershed or aquifer recharge area that have a potential to contaminate water.

contaminant Anything found in water other than hydrogen or oxygen.

contamination Any introduction into water of microorganisms, chemicals, wastes, or wastewater in a concentration that makes the water unfit for its intended use.

contingency plan A document that details the action of a water utility under specified adverse conditions.

corrosion Destruction by chemical action.

curbs Precast concrete liners for dug wells.

cyst A resistant form of a living organism.

daily flow The total volume of water (in gallons or liters) passing through a plant during a 24-hour period.

demand factor The ratio of the peak or minimum demand to the average demand.

demand management A conservation term that refers to controlling the users' use of water.

density stratification The formation of layers of water in a reservoir, with the water that is the densest at the bottom and least dense at the surface.

detention time The period for which water is held before some further action takes place.

disinfection The water treatment process that kills disease-causing organisms in water, usually by adding chlorine.

dissolved oxygen (DO) The oxygen dissolved in water, wastewater, or other liquid, usually expressed in milligrams per liter, parts per million, or percent of saturation.

dissolved solids Any material that is dissolved in water and can be recovered by evaporating the water after filtering the suspended material.

divide The line that follows the ridges or summits forming the boundary of a drainage basin (watershed) and that separates one drainage basin from another. Also called *watershed divide*.

drainage basin An area from which surface runoff is carried away by a single drainage system. Also called *catchment area* or *watershed drainage area*.

drawdown The difference between the static water level and the pumping water level in a well.

drawdown method A testing procedure to determine the characteristics of an aquifer or well.

dual system A double system of pipelines, one carrying potable water and the other water of lesser quality.

E. coli A bacterial species; one of the fecal coliforms.

effluent Water flowing out of a structure such as a treatment plant.

electrical conductivity (EC) A test that measures the ability of water to transmit electricity. Electrical conductivity is an indicator of dissolved solids concentration. Normally an EC of 1,000 μmhos per square centimeter indicates a dissolved solids concentration of 600–700 mg/L.

eminent domain The power to seize property.

ephemeral stream (1) A stream that flows only in direct response to precipitation. Such a stream receives no water from springs and no continued supply from melting snow or other surface source. Its channel is above the water table at all times. (2) Streams or stretches of streams that do not flow continuously during periods of as much as 1 month.

eutrophication A process by which a lake becomes too rich in aquatic life.

evaporation The process by which water becomes a vapor at a temperature below the boiling point. The rate of evaporation is generally expressed in inches or centimeters per day, month, or year.

extension agents Government officials who serve as consultants to agriculture.

fecal coliform A bacterial group that is an indicator of human or animal pollution.

filterable residue test A test used to measure the total dissolved solids in water by first filtering out any undissolved solids and then evaporating the filtered water to dryness. The residue that remains is called *filterable residue* or *total dissolved solids.*

filtration The process of removing solids from water by passing the water through some material or device.

flotation A process for separating solids from water by using air to float the particles.

flow General term for movement of water, commonly used to mean (imprecisely) instantaneous flow rate, average flow rate, or volume.

flow rate The volume of water passing by a point per unit of time. Flow rates are either instantaneous or average.

frazil ice Small ice crystals that can block water intakes.

free residual chlorine Hypochlorous acid in water that responds to testing procedures and that serves as an active disinfectant.

geographic information system (GIS) A computer hardware and software system that captures, analyzes, and displays interrelated and geographically linked data.

Giardia lamblia A protozoan that can survive in water and that causes human disease.

granular activated carbon A chemical used to remove certain dissolved contaminants from water.

granular media A material used for filtering water, consisting of grains of sand or other material.

gravel pack Gravel surrounding the well intake screen, artificially placed ("packed") to aid the screen in filtering out the sand of an aquifer. Needed only in aquifers containing a large proportion of fine-grained material.

groundwater Subsurface water occupying the saturation zone, from which wells and springs are fed. In a strict sense, the term applies only to water below the water table.

groundwater aquifer See *aquifer*.

groundwater classification A scheme to categorize groundwater according to its quality.

hardness A characteristic of water, caused primarily by the salts of calcium and magnesium. Causes deposition of scale in boilers, damage in some industrial processes, and sometimes objectionable taste. May also decrease the effectiveness of soap.

hazardous-waste facilities Depository areas where dangerous materials are stored.

hydraulic conductivity A measure of the ease with which water will flow through geologic formations.

hydraulic jetting The use of water forced through the well screen to suspend fine particles in well development.

hydraulics The branch of science that deals with fluids at rest and in motion.

hydrodynamics The study of water in motion.

hydrologic cycle The water cycle; the movement of water to and from the surface of the earth.

hydrostatics The study of water at rest.

impermeable layer A layer not allowing, or allowing only with great difficulty, the movement of water.

impervious Resistant to the passage of water.

impoundment A pond, lake, tank, basin, or other space, either natural or constructed, that is used for storage, regulation, and control of water.

indicator organisms A bacterium that does not cause disease but indicates that disease-causing bacteria may be present. See *fecal coliform*.

indirect potable reuse The use of water from streams that have upstream discharges of wastewater.

industrial water use Use of water in industrial activities such as power generation, steel manufacturing, pulp and paper processing, and food processing.

infiltration The flow or movement of water through soil. Also called *percolation*.

infiltration gallery A subsurface structure to receive water filtered through a stream bed.

inorganic material Chemical substances of mineral origin, or more correctly, not of basically carbon structure.

instantaneous flow rate The flow rate (volume per time) at any instant. Defined by the equation $Q = AV$, where Q = flow rate, A = cross-sectional area of the water, and V = velocity.

intake structure A structure or device placed in a surface water source to permit the withdrawal of water from that source.

ion exchange A process that removes undesirable chemicals from water and replaces them with harmless ones.

maximum contaminant level (MCL) The maximum permissible level of a contaminant in water as specified in the regulations of the Safe Drinking Water Act.

membrane processes Water purification processes using thin plastic films.

milligrams per liter (mg/L) A unit of the concentration of water or waste-water constituents. One mg/L is equivalent to 0.001 g of the constituent in 1,000 mL (1 L) of water. For reporting the results of water and wastewater analyses, this unit has replaced parts per million, to which it is approximately equivalent.

minimum day The day during which the minimum-day demand occurs.

minimum-day demand Least volume per day flowing through the plant for any day of the year.

minimum-hour demand Least volume per hour flowing through a plant for any hour in the year.

minimum-month demand Least volume of water passing through the plant during a calendar month.

molecule The smallest particle of a compound that has the compound's characteristics.

mrem Millirem. See *rem*.

National Pollution Discharge Elimination System (NPDES) A licensing system for waste discharges.

nephelometric turbidity unit (ntu) The unit of measure used to express the turbidity (cloudiness) of a water sample.

nitrogen fertilizer A plant nutrient that can cause algae blooms in sensitive water bodies.

nonpoint sources Material entering a water body that comes from overland flow rather than out of a pipe.

numerical modeling A computerized technique for determining an aquifer recharge area.

observation well Well placed near a production well to monitor changes in the aquifer.

organic adsorption The removal of volatile organic chemicals and synthetic organic chemicals by use of activated carbon.

organic material Chemicals that contain carbon and are associated with living matter.

overlay maps Transparent sheets containing data that are placed over the base map and indicate special features.

overlying use The land use that occurs on top of an aquifer.

oxidant A chemical that oxidizes other materials.

oxidize To chemically combine with oxygen, or in the case of bacteria or other organics, the combination of oxygen and organic material to form more stable organics or minerals.

oxygen depletion A state in which oxygen has been used up so that little or none is left.

ozone A form of oxygen that has very strong oxidizing powers.

palatable Pleasing to the taste.

parts per million (ppm) Unit of measure indicating the number of weight or volume units of a constituent present for each 1 million units of the solution or mixture. Formerly used to express the results of most water and wastewater analyses, but recently replaced by the use of milligrams per liter.

pathogens Disease-causing organisms.

peak day The day during which the peak-day demand occurs.

peak-day demand The greatest volume per day flowing through a plant for any day of the year.

peak-hour demand The greatest volume per hour flowing through a plant for any hour in the year.

peak-month demand The greatest volume of water passing through the plant during a calendar month.

percolation The movement or flow of water through the pores of soil, usually downward.

perennial stream A stream that flows continuously at all seasons of a year and during dry as well as wet years. Such a stream is usually fed by groundwater, and its water surface generally stands at a lower level than that of the water table in the locality.

pH A measure of the acidic or alkaline nature of water. Values of pH range from 1 (most acidic) to 14 (most alkaline); pure water, which is neutral (neither acidic nor alkaline), has a pH of 7, the center of the range.

phosphorus fertilizer A plant nutrient that can cause algae blooms in sensitive water bodies.

photochemical oxidation Changes in chemicals caused by sunlight.

photosynthesis The process by which plants, using the chemical chlorophyll, convert the energy of the sun into food energy. Through photosynthesis, all plants and ultimately all animals (which feed on plants or other, plant-eating animals) obtain the energy of life from sunlight.

piezometric surface The surface that coincides with the static water level in an artesian aquifer.

point source Wastewater coming from a discharge pipe.

police power The ability to enforce laws.

pollution A condition created by the presence of harmful or objectionable material in water.

poorly buffered Not resisting changes in pH.

porosity An indication of the volume of space within a given amount of a material.

powdered activated carbon Carbon, added to water in powdered form, used to remove chemicals from water.

precipitation The process by which atmospheric moisture is discharged onto land or water surface. Precipitation, in the form of rain, snow, hail, or sleet, is usually expressed as depth for a day, month, or year, and designated as daily, monthly, or annual precipitation.

primacy The acceptance by states of the task of enforcing the SDWA.

priority use The assigning of water rights based on who has been using the water the longest.

pristine Pure and free from contamination.

protozoa Small single-celled animals including amoebae, ciliates, and flagellates.

public water use Use of water to support public facilities such as public buildings and offices, parks, golf courses, picnic grounds, campgrounds, and ornamental fountains.

quagga mussel A bivalve that multiplies rapidly in fresh water and can clog intake pipes.

qualified right A water rights term indicating that the right to water use is not absolute but must be shared with others.

radial well A very wide, relatively shallow caisson that has horizontally drilled wells with screen points at the bottom. Radial wells are large producers.

radioactive Referring to a material that has an unstable atomic nucleus, which spontaneously decays or disintegrates, producing radiation.

rainfall intensity The amount of rainfall occurring during a specified time. Rainfall intensity is usually expressed as inches or centimeters per hour.

reasonable use A water rights term indicating that the water use is acceptable in general terms.

recharge Addition of water to the groundwater supply from precipitation and by infiltration from surface streams, lakes, reservoirs, and snowmelt.

recovery method A procedure for aquifer evaluation that measures how quickly water levels return to normal after pumping.

rem A unit of measure of radioactive effects on humans.

reuse water Wastewater treated to make it useful.

riparian doctrine A water right that allows the owners of land abutting a stream or other natural body of water to use that water.

riparians   Those whose property holdings are along the shores of a water body.

rule of correlative rights   A water rights rule specifying that rights are not absolute but depend also on the rights of others.

safe yield   The maximum dependable water supply that can be withdrawn continuously from a surface water or groundwater supply during a period of years in which the driest period or period of greatest deficiency in water supply is likely to occur.

saltwater intrusion   The invasion of an aquifer by salt water because of overpumping of a well.

sand barrier   A layer of gravel around the curb of a dug well.

sanitary survey   A detailed inspection of a facility with a view to public health concerns.

saturated   The condition of a material when it can absorb no more of a second material. Saturated soil has its void spaces completely filled with water, so any water added will run off and not soak in.

screens   A device for excluding debris on a surface water intake and for excluding sand in a groundwater supply.

SDWA   The Safe Drinking Water Act.

secondary maximum contaminant level (SMCL)   A nonenforceable regulation on a nonhealth-related contaminant.

sedimentation   A treatment process using gravity to remove coagulated particles.

seepage   The slow movement of water through small cracks or pores of a material.

service outlet   A device used for releasing water at a dam for downstream uses.

siltation   The accumulation of silt (small soil particles between 0.00016 and 0.0024 in. [0.004 and 0.061 mm] in diameter) in an impoundment.

slide gate   A device on a dam to control the release of water.

sole-source aquifer   The single water supply available in an area; also a USEPA designation for water protection.

solubilization   The process of making a material easier to dissolve.

specific-capacity method A testing method for determining the adequacy of an aquifer or well.

specific yield A measure of well yield per unit of drawdown.

spillway A device used to release water from a dam.

spring A location where groundwater emerges on the surface of the ground.

sterilization The removal of all life from water, as contrasted with *disinfection.*

stream bed permeability Measure of the ability of the bottom of a watercourse to transmit water to and from the underground.

stream channel The course through which a stream flows.

supply management The implementation of techniques by a utility to save water.

surface runoff (1) That portion of the runoff of a drainage basin that has not percolated beneath the surface after precipitation. (2) The water that reaches a stream by traveling over the soil surface or by falling directly into the stream channels, including not only the large permanent streams but also the tiny rills and rivulets.

surface water All water on the surface, as distinguished from *groundwater.*

surger A device used to develop a well.

surging and bailing A method used to develop a well by alternately increasing and decreasing water pressure against the walls of the well to dislodge small particles.

synthetic organic chemicals (SOCs) Chemicals produced by humans that can be water contaminants.

temperature A physical characteristic of water. The temperature of water is normally measured on one of two scales: Fahrenheit or Celsius.

thermal stratification The layering of water as a result of temperature differences.

thermoelectric power Electricity produced by heat in which water is used for steam production.

threshold odor The minimum odor of a water sample that can just be detected after successive dilutions of odorless water.

threshold odor number (TON) A numerical designation of the intensity of odor in water.

time of travel The time required for groundwater recharge to percolate through the soil and get to the wellhead.

topographic map A map that shows the elevation of the land in a specified area.

transpiration The process by which water vapor is lost to the atmosphere from living plants.

trihalomethanes (THMs) Compounds of chlorine and organic matter that are generated during disinfection.

turbidity A physical characteristic of water making the water appear cloudy. The condition is caused by the presence of suspended matter.

turnover The vertical circulation of water in large water bodies caused by the mixing effects of temperature changes and wind.

ultraviolet irradiation Use of a portion of the light spectrum to disinfect drinking water.

unaccounted-for water Water use that does not go through meters and thus is not paid for (including water that is wasted).

viruses The smallest and simplest form of life. The many types of viruses reproduce themselves in a manner that causes infectious disease in some larger life forms, such as humans.

void spaces A pore or open space in rock or granular material that is not occupied by solid matter. It may be occupied by air, water, or other gaseous or liquid material.

volatile organic chemicals (VOCs) Organics that are produced by humans, are primarily solvents, and are found in water as contaminants.

waiver for filtration An exception granted to a state so that a water source can be served without prior filtration.

water body A natural or artificial basin in which water is impounded.

waterborne disease A disease caused by a waterborne organism or toxic substance.

water conservation The reduction of water usage, accomplished primarily by preventing waste.

watercourse (1) A running stream of water. (2) A natural or artificial channel for the passage of water.

water rights A body of law that determines water ownership.

water-saving fixtures   Devices and appliances specifically designed to have low water use.

water-saving kits   Devices sometimes distributed by utilities to reduce customer water usage.

watershed   See *drainage basin*.

watershed boundary   The watershed divide (the high point on a watershed), which directs the flow of water in a certain direction.

water table   The upper surface of the zone of saturation closest to the ground surface.

water-table aquifer   An aquifer confined only by a lower impermeable layer.

water-table well   A well constructed in a water-table aquifer.

watershed divide   See *watershed boundary*.

well development   The process of removing particles from an aquifer to produce potable water.

well field placement   The arranging of wells to achieve a desired output from an aquifer.

wellhead protection   A process of controlling potential groundwater contamination for a specific groundwater source.

wellhead protection program (WHPP)   A program required by the Safe Drinking Water Act to achieve wellhead protection.

Xeriscaping   The selection and use of trees, shrubs, and plants that require small amounts of water for landscaping.

zebra mussel   A bivalve that has become a serious pest in surface water because it blocks intake pipes.

zone of saturation   The part of an aquifer that has all the water it can contain.

zoning   A civil process that controls land use.

INDEX